RANGE ECONOMICS

BIOLOGICAL RESOURCE MANAGEMENT

A Series of Primers on the Conservation and Exploitation of Natural and Cultivated Ecosystems

Wayne M. Getz, Series Editor
University of California, Berkeley

Available
Range Economics, by John P. Workman

Forthcoming in 1986
Adaptive Management of Renewable Resources, by Carl Walters
Building Models for Wildlife Management, by Anthony Starfield and A. L. Bleloch
Mathematical Programming for Economic Analysis in Agriculture, by Peter B. R. Hazell and Roger D. Norton

RANGE ECONOMICS

JOHN P. WORKMAN

Department of Range Science
Utah State University

MACMILLAN PUBLISHING COMPANY
NEW YORK

Collier Macmillan Publishers
LONDON

Copyright © 1986 by Macmillan Publishing Company
A division of Macmillan, Inc.

All rights reserved. No part of this book may be reproduced or transmitted in any form or by any means, electronic or mechanical, including photocopying, recording, or by any information storage and retrieval system, without permission in writing from the Publisher.

Macmillan Publishing Company
866 Third Avenue, New York, NY 10022

Collier Macmillan Canada, Inc.

Printed in the United States of America

printing number
1 2 3 4 5 6 7 8 9 10

Library of Congress Cataloging in Publication Data

Workman, John P.
    Range Economics

    (Biological resource management)
    Includes bibliographies and index.
    1. Rangelands—Economic Aspects.   2. Range management.   I. Title. II. Series.
HD1635.W67 1986   333.74   85-18882
ISBN 0-02-948810-9

Dedicated to my Family

Contents

Foreword	vii
Preface	ix
CHAPTER 1. AN OVERVIEW OF RANGE ECONOMICS	**1**
Range Economics Defined	1
Range Economics Compared with Ranch Economics and Range Management	2
Historic Sketch of Range Economics	3
Purpose and Content of This Book	5
Literature Cited	6
CHAPTER 2. ECONOMICS IN A RANGE MANAGEMENT SETTING	**10**
Importance of Economics to the Range Manager	10
Setting the Stage—Why Rangelands are Grazed	12
The Western Cattle Ranch—A Sketch of the Operation	13

The Standard Ranch Income Statement	15
The Modified Ranch Income Statement	18
The Ranch Financial Statement	20
Literature Cited	22

CHAPTER 3. SUPPLY AND DEMAND—THE BASIC TOOLS 25

Demand	25
Supply	27
Market Equilibrium	29
Curve Shifts versus Movements Along Curves	30
Demand Shifters	32
Supply Shifters	32
An Application	33
Appendix I. Elasticity of Demand for Beef and Its Effect on Range Improvement Decisions	34
Literature Cited	37

CHAPTER 4. THE PRODUCTION FUNCTION 38

Production Economics—The Basic Problems	38
Theory of the Production Function	39
Law of Diminishing Returns	40
Stages of Production	41
Stage II—The Relevant Stage	42
Appendix I. The General Production Function Model	45
Further Reading	46

CHAPTER 5. OPTIMUM INTENSITY OF PRODUCTION 48

The Marginality Principle	48
Marginal Revenue	50
Fixed versus Variable Costs	50
Marginal Cost	51
Equating at the Margin for Optimum Intensity	51
Two Common Fallacies	52
Factors Affecting Optimum Intensity and Profit	56
The General Marginality Model	61
Total Revenue Minus Total Cost Approach	61
MC = MR Approach	64
The Value of Marginal Product (VMP) Approach	73
Literature Cited	79
Further Reading	80

CONTENTS ix

CHAPTER 6. OPTIMUM COMBINATION OF INPUTS 85

A Tabular Production Function 85
A Graphical Example 86
   The Minimum Cost Combination 87
   The Maximum Product Combination 88
   The Optimum Production Level 88
The General Isoquant–Isocost Model 89
   Derivation of the Isoquant Map 89
   Characteristics of Isoquants 91
   Ridge Lines—The Relevant Stage 92
   The Isocost Line and the Optimum Mix 94
   The Expansion Path and the Optimum Production Level 95
   Input Adjustments to Achieve Optimum Production 97
   The Numerous Variable Input Case 99
Literature Cited 100

CHAPTER 7. OPTIMUM COMBINATION OF OUTPUTS— THE ECONOMICS OF MULTIPLE USE 102

Competitive Products 103
   Competition at Constant Rates of Substitution 103
   Competition at Increasing Rates of Substitution 106
Complementary Products 110
   Biological Interpretation of Complementary Products 111
   Economic Interpretation of Complementary Products 111
   Calculating Nonmarket Prices Justifying Allocation Decisions 111
Supplementary Products 112
   Biological Interpretation of Supplementary Products 113
   Economic Interpretation of Supplementary Products 114
Antagonistic Products 114
   Biological Interpretation of Antagonistic Products 115
   Economic Interpretation of Antagonistic Products 116
Literature Cited 116

CHAPTER 8. DISCOUNTING—ADJUSTING COSTS AND RETURNS FOR THE EFFECTS OF TIME 118

Why Does It Cost to Wait? 118
Compounding 120
Discounting 124
Flow of Costs and Returns 125
Capitalization 131
Sinking Fund 133

Future Worth of Annual Deposits 135
Credit Decisions—Where to Borrow Funds 138
Literature Cited 139

CHAPTER 9. ECONOMIC ANALYSIS OF PRIVATE RANGE IMPROVEMENTS 141

The Analysis—A General Overview 141
   Present Net Worth 142
   Benefit–Cost Ratio 142
   Internal Rate of Return 143
Information Required for Analysis 143
   Expected Project Benefits 144
   Value of Expected Benefits 146
   Expected Projected Life 156
   Expected Project Costs 156
   Interest Rate on Borrowed Capital 158
Analysis and Interpretation 159
   Forage Value Based on Private Lease Rate 159
   Forage Value Based on Livestock Production 161
Appendix I. Disagreement among Investment Criteria—
   A Proposed Solution 163
   A Review of the Problem 163
   Some Range Improvement Examples 164
   Normalization—A Solution 167
   Selecting an Optimum Combination from Three or More Projects 172
Appendix II. Modified IRR Calculations for Nonuniform Flows
   of Costs and Returns 174
   The Concept 174
   An Application 175
Literature Cited 177

CHAPTER 10. BENEFIT–COST ANALYSIS ON PUBLIC RANGES 183

Role of Economic Analysis on Public Lands 183
   Analysis versus Justification 184
   The Ravages of Time 185
Economic Feasibility and Public Range Investment 186
   An Idealized Benefit–Cost Analysis 187
   Value Comparison Problems 189
Cost Effectiveness and Nonmarket Pricing 190
   Cost Effectiveness: A "Specified Bang for the Smallest Buck" 190
   Nonmarket Pricing: A Search for Values 192

Cost Effectiveness and Nonmarket Pricing—Summary	197
The Importance of the Interest Rate Used for Discounting	198
Appendix I. Selecting the "Correct" Discount Rate	200
Opportunity Cost versus Social Discount Rate	201
Adjusting for Risk	203
Adjusting for Inflation	204
Literature Cited	206

INDEX 211

Foreword

The range livestock industry is a highly complex and diversified industry. The economic performance and well-being of ranchers varies considerably. Thus, no single indicator of performance can capture the variation that exists among ranches and among range managers. Ranges have an array of physical influences, including topography, climate, type of vegetation, soils, slope, and range conditions. Ranches, too, have an array of economic and social conditions, including managerial ability, length of planning horizons, and motivation for ranching. The financial position of ranchers varies from outright ownership to servicing considerable debt. One conclusion is obvious: the Western range livestock industry is not homogeneous. It is dynamic in nature, with monetary returns, factor costs, forage production, and net returns varying greatly with time.

This book is intended as a guide in performing appropriate economic analysis of range improvements and management practices. The student of range economics is introduced to issues confronting every land manager, whether operating on public or private land areas. Real-world examples are presented which clarify current rangeland problems and pragmatically outline the economic logic necessary to arrive at a correct economic solution. Professional managers are exposed to tools allowing them to make their own determination of the economic consequences of specific management decisions.

Economic justification is now a part of virtually every regulation, policy, and enactment pertaining to rangeland. The terms *cost-effective*, *favorable benefit/cost ratios*, and *net present value* are heard with increasing regularity. No longer are rangeland managers looking at problems associated with a single output at a single point in time; rather, they are confronted with problems encompassing multiple uses over a period of time. This text develops the tools necessary for sound rangeland stewardship under competing demands. The concepts and applied examples presented in this text will afford the necessary foundation for the rangeland manager to make an informed decision in increasingly complex issues.

> John M. Fowler
> New Mexico State University

Preface

This book was written in response to a need perceived by the Society for Range Management (SRM) for a range economics textbook for college seniors and professional range managers. In 1973, SRM President Dr. Floyd Kinsinger had the foresight to appoint an SRM Rangeland Economics Sciential Committee. Four years later, at the urging of SRM and even more personally convinced of the need for a range economics textbook, I agreed to develop a draft of such a book and to test it on range management students at Utah State University. Early versions of this book have been read by students since 1978.

This book is intended for use as a guide in performing economic analyses of range improvements and management practices. It is not meant to be a review of range economics research results nor a catalog of range management practices expected to prove profitable. Usefulness of these latter approaches would be very short-lived due to changes in prices of products and inputs. SRM originally requested a state-of-the-art summary of established economics principles and concepts relevant to range management. This book attempts such a summary and, hopefully, also provides responses to questions often asked by range management students and professionals.

The contributions of several individuals toward the completion of the book deserve acknowledgment. I am particularly grateful to two of my former teachers: Dr. Jack Hooper for sparking my interest in range economics and Dr. Del Gardner for sustaining it. I thank Dr. Don Dwyer for recommending not one, but two, sabbatical leaves to work on this book. I am grateful to Dr. Thad Box for his advice and encouragement and especially for his belief in the book from the beginning. I extend a special thanks to Dr. John Fowler for his helpful technical review of the final draft.

Two institutions deserve special mention: Winrock International Livestock Research and Training Center, Petit Jean Mountain, Arkansas, provided the housing, travel support, office, typing services, and creative atmosphere that allowed completion of the first draft. The Department of Animal and Range Sciences at New Mexico State University, Las Cruces, furnished the quiet office and Sun Country winter conducive to finishing the final draft.

Finally, my thanks to Lou for lending her support to this and numerous other questionable projects during our 21 years together.

John P. Workman
Utah State University

RANGE ECONOMICS

Chapter 1

An Overview of Range Economics

RANGE ECONOMICS DEFINED

Range economics is best defined, perhaps, by providing separate definitions of the terms range and economics. Definitions of *range* and its synonym *rangeland* are numerous and varied. They include short, simple definitions such as "...range is nontillable land whose major use is the production of forages" (Gray, 1968) and more complex and exacting definitions such as "rangelands are those areas...which by reason of physical limitations—low precipitation, rough topography, poor drainage, or cold temperatures are unsuited to cultivation...are a source of forage for free-ranging native and domestic animals as well as a source of wood products, water, and wildlife" (Stoddart, Smith, and Box, 1975). General agreement seemingly found among the various definitions is that range is a specific kind of land rather than land devoted to a specific kind of use (livestock grazing).

The term economics is also blessed with a wealth of definitions (Samuelson, 1964), but for the purpose at hand most economists would probably accept the following definition. Economics is the science that deals with efficient allocation of scarce resources among competing uses. The resources may be natural (such as rangeland and oil), human (such as labor and management), or man-made (such as fencing materials and pickup trucks). The competing uses might be sheep competing with cattle for forage, grazing competing with rural residences for land, or windmills competing with campsites for development capital.

Range economics is the science of applying the principles of economics and range management simultaneously to determine the economic consequences of decisions involving the use, development and/or protection of rangelands. It is difficult to determine just when range economics became a separate discipline distinguishable from its range management and agricultural economics roots. Since the mid-1950s interest and effort in range economics has alternately waxed and waned, but at any point during that time span there have probably been no less than 10 and no more than 40 professional people affiliated with universities or land management and research agencies of the federal government who claimed the title of "range economist."

Some agricultural economists might argue that the distinction between a range economist and any other type of agricultural, resource, or production economist is an artificial one and that any trained economist could do the job that range economists claim to do. Most range managers more readily accept range economics as a distinguishable discipline. There seems to be some agreement among range managers that it may be more efficient to assign the economic problems of range use and improvement to scientists trained in both economics and range science rather than attempting to solve such problems through interdisciplinary study teams composed of traditional agricultural economists and range managers who lack training in economics. However, it is only fair to acknowledge that many of the important research contributions in range economics have been made by economists who had no formal training in range management but who enlisted the help of range scientists in designing experiments and interpreting results.

In summary, range economics is a comparatively young discipline and one that is still experiencing difficulties in establishing an identity.

RANGE ECONOMICS COMPARED WITH RANCH ECONOMICS AND RANGE MANAGEMENT

In contrast to the youth of range economics, *ranch economics* studies were performed as early as shortly after the Civil War. Ranch economics has been defined by Gray (1968) as the "application of economics to ranch management decisions" and differs from range economics primarily in scope. Ranch economics deals with the economics of management decisions from the viewpoint of the ranch as a business unit, while range economics focuses more broadly on the economic consequences of range management decisions on lands that may be in private or public ownership (or both) from both the viewpoints of the individual ranch unit or the public land management agency or a combination of the two. Range economics grew out of the combined efforts of range scientists and agricultural and resource economists to solve mutual problems involved with the use of mixtures of private and public rangeland resources lying primarily in the 11 western states. The problems faced and conceptual framework developed have application to a large portion of the rangelands of the world since combinations of private and public land and dicotomies in management goals are almost universal.

Range management as a scientific discipline emerged between 1910 and 1920, about 50 years after ranch economics and nearly 50 years before the beginnings of range economics as it is defined here. Range management has been defined in the most recent range management textbooks as "...the science and art of optimizing the returns from rangelands in those combinations most desired by and suitable to society through the manipulation of range ecosystems" (Stoddart, Smith, and Box, 1975) and as "...the discipline that...applies...range science to natural resource systems for...protection, improvement, and continued welfare of the basic range resource...and optimum production of goods and services in combinations needed by mankind" (Heady, 1975). The two definitions are strikingly similar. Both emphasize that rangelands produce a variety of benefits in addition to livestock products and that optimum levels and mixtures of the various products are important goals. Both definitions also imply that the rangeland resource is at least partially owned by the public and therefore at least part of the total rangeland products should accrue to society in general. The further implication, then, is that range management is a social science, serving the needs of society, as well as a biological science concerned with perpetuation of the basic range resource. It may be worth noting that both of these definitions of range management are oriented much more towards multiple use and the social aspects of range management than were the definitions appearing the first two editions of the classic *Range Management* by Stoddart and Smith (1943, 1955). Range economics, then, while being broader in scope than ranch economics, is much narrower than range management. Range management encompasses all aspects of the rangeland resource including all of the various ways this land might be used and all the numerous disciplines (biology, ecology, sociology, soils, meteorology, etc.) required to adequately study and plan its use. Range economics is one of several subdisciplines of range management and is confined to determination of the economic consequences of range management decisions. Thus, range economics relies on the various other subdisciplines within range management to provide information concerning what is physically and biologically possible. One last comparison that may be useful is that range managers deal primarily with questions of "will it work," while range economists focus on questions concerning "will it pay" (Subcommittee on Range Research Methods of the Agricultural Board, 1962).

HISTORIC SKETCH OF RANGE ECONOMICS

It is difficult to pinpoint the beginnings of range economics as a distinct discipline. Studies in ranch economics in Nebraska were completed as early as 1910 by Smith. Morton's production cost study appeared in 1914. In 1928 information concerning Utah sheep ranch economics was published by Esplin et al. and costs of Nevada cattle operations were reported by Brennan and Smith. Published treatments of ranch economics in book form came soon after, notably Osgood's (1929) *The Day of the Cattleman* and Dale's (1930) *The Range Cattle Industry*. Nearly 20 years later the textbook *Farm Management* (Black et al., 1947) appeared. It was devoted in part to the economics and management of western livestock ranches. The year 1950 was a notable one in that two books treating range livestock ranching were published: Saunderson's *Western Stock Ranching* and Clawson's widely read *The Western Range Livestock Industry*. The latter book was perhaps the first successful attempt to provide not only a detailed summary of the current and historical economic status of range livestock operations but a treatment of range improvements, public land grazing, and multiple use of rangelands as well.

The economics of range reseeding was first analyzed by Pearse and Hull in 1943. The Bureau of Agricultural Economics (BAE) report on range reseeding appeared in 1949. During this entire period the BAE remained active in ranch economics research (Hockmuth and Goodsell, 1948) and this agency later became the Economic Research Service (Goodsell and Gray, 1961).

In 1952 another textbook devoted partly to the economic aspects of range livestock grazing was published, Ciracy-Wantrup's *Resource Conservation, Economics, and Policies*. Perhaps even more notable from the standpoint of the development of range economics as a discipline was the appearance of what might be considered the first economics paper in the *Journal of Range Management* (Hockmuth, 1952). One year later, a discussion of the economics of range improvements appeared in the *Journal of Farm Economics* (Upchurch, 1953).

Hopkin's (1954) dissertation at Iowa State University entitled "Economics of Western Range Resource Use" was probably the first Ph.D. dissertation completed in range economics, followed by those of Lloyd (1959) at Utah State University, McConnen (1960) at University of California, Berkeley, Nielsen (1965) and Bromley (1968) both at Oregon State University, and Hooper (1969) at University of California, Berkeley. Spanning roughly this same period (1951–1968), the Western Agricultural Economics Research Council Committee on Economics of Range Use and Development devoted what was undoubtedly the most extensive committee effort ever brought to bear on range economics topics. The committee met annually to work on current economic problems in management and improvement of rangeland and a series of 12 mimeographed reports was published between 1957 and 1968 (Baker and Plath, 1957; McCorkle, 1959; Fulcher and Brough, 1961; Roberts and Gray, 1961; Roberts, 1962, 1963, 1964; Wennergren, 1965, 1966; Nielsen, 1967; Lansford and Gray, 1968; and Gray, 1969). Unfortunately, the reports did not receive widespread distribution outside the Western Agricultural Economics Association and little of this wealth of information reached the hands of range management professionals. This series of reports, some addressing topics relevant primarily to the time written and others timeless, is highly recommended for anyone interested in range economics, whatever their training or experience.

Meanwhile, two important publications appeared—one a summary dealing with the design and interpretation of research in range economics (Subcommittee on Range Research Methods of the Agricultural Board, 1962) and the other a guide to aid ranchers in performing their own economic analyses of range improvements (McCorkle and Caton, 1962).

Gray et al. (1965), writing from the viewpoint of the ranch owner, summarized the expected economic results of a wide variety of range improvements in the Southwest. In 1967, Nielsen used a Utah operation to develop a guide for ranchers in making economic analyses of range improvements.

The year 1968 marked the appearance of a major publication in the economics of the private range livestock operation: Gray's widely used textbook *Ranch Economics*. The Western Agricultural Economics Research Council (WAERC) range committee was disbanded the next year after publication of the last in the series of reports (Gray, 1969). No summary or state-of-the-art publication has been written since 1969 except for *Economics and Management Planning of Range Ecosystems* (Jameson et al., 1974), a textbook that provided good coverage of linear programming and more complex stochastic and recursive programming applicable to range management decisions. Individuals and small teams of authors continued to publish range economics research results in the form of Economic Research Service reports, agricultural experiment station bulletins, and papers in the *Journal of Range Management*, *Western Journal of Agricultural Economics*, and *American Journal of Agricultural Economics*. But there was no further work by widely represented committees of the WAERC type until 1982, when a Range Economics Symposium was held (Wagstaff, 1983). It was the consensus of this group that a regional range research coordinating committee would be desirable.

Included in the proceedings of the 1982 Range Economics Symposium was a history of range economics research by Gray (1983) in which he divided the time span 1870–1982 into the (1) frontier, (2) descriptive, (3) renaissance, and (4) analytical–public policy periods. Gray's history is recommended for anyone interested in range economics research.

PURPOSE AND CONTENT OF THIS BOOK

This book is intended for use as a guide to making economic analyses of range management and improvement practices. Hopefully, it will provide the conceptual framework necessary for the professional range manager to make his or her own determination of the economic consequences of a particular management decision. The book does not attempt to provide a list of range improvement and management practices that can be expected to prove profitable for the private ranch owner or economically feasible for the manager of public rangeland. While such an approach might be welcomed by range managers who would prefer not to spend their limited time doing economic analyses, the dynamics of input and product price relationships would restrict the usefulness of such an approach to a very short time period. Instead the book attempts to equip the range manager with techniques that enable economic evaluation tailored to local biological production and price situations.

The book is directed primarily to range management students currently enrolled in university classes and to practicing range managers, including federal and state land managers, range management consultants, college and university teachers, researchers, and ranchers. It is designed to serve as a textbook for a three semester hour course in range economics for senior range management students. Hopefully, it will also serve as an understandable source for professionals working in range management and as a flexible reference for range economics short courses and workshops. Mathematical footnotes are included for range economics graduate students and other readers with more advanced mathematical backgrounds.

The book consists of 10 chapters. Chapter 2 introduces range economics as a necessary tool in range management against a backdrop of the day-to-day workings and economic problems of western ranch operations, an important discussion in view of the shrinking proportion of range management students who possess practical farm or ranch experience. The fundamental workings of supply, demand, and price in the competitive market are reviewed in Chapter 3, using rangeland products as examples. Chapter 4 explores the biological production function, the basic relationship upon which all decisions involving land use intensity rest. Chapter 5 focuses on traditional single input–single output production economics theory from a range management viewpoint. Application of marginal analysis to the problem of determining correct grazing intensity is developed in detail. Chapter 6 expands the production economics principles to include the case of the optimum mix of two or more inputs which are substitutes in rangeland production. In Chapter 7 the analysis is enlarged to include the important problems of multiple use—the case where a tract of land is biologically capable of producing two or more products and a decision must be made concerning the correct product mix. Chapter 8 deals with the crucial concept of time comparisons that underlies all capital investment and borrowing decisions. Chapter 9 develops the details of performing economic analyses of private land range improvements. Finally, Chapter 10 recognizes that the goals of the public range manager are sometimes quite different than those of his private land counterpart and alterations in economic analysis necessary for public land decisions are discussed.

LITERATURE CITED

Baker, C. B. and C. V. Plath (Eds.). 1957. *Economic Research in the Use and Development of Range Resources—A Methodological Anthology*. Western Agricultural Economics Research Council, Berkeley, CA. 151 pp. (mimeo.).

Black, J. D., M. Clawson, C. R. Sayre, and W. W. Wilcox. 1947. *Farm Management*.

Macmillan, New York. 1073 pp.
Boykin, C. C. and R. J. Hildreth. 1958. Management aspects of range management. *J. Range Manage.* 11:173–176.
Brennan, C. A. and G. H. Smith. 1928. *Preliminary Report on a Study of Cattle Production in Nevada*. Nevada Agricultural Experiment Station Bulletin No. 111. 14 pp.
Bromley, D. W. 1968. Economic importance of federal grazing: an interindustry analysis. Ph.D. dissertation, Oregon State University, Corvallis.
Bureau of Agricultural Economics. 1949. *Will More Forage Pay*. U.S. Department of Agriculture miscellaneous publication No. 702. 90 pp.
Ciracy-Wantrup, S. V. 1952. *Resource Conservation, Economics, and Policies*. Univ. California Press, Berkeley.
Clawson, M. 1950. *The Western Range Livestock Industry*. 1st ed. McGraw-Hill, New York. 401 pp.
Dale, E. E. 1930. *The Range Cattle Industry*. Univ. of Oklahoma Press. Norman, OK.
Esplin, A. C., W. Peterson, P. V. Cardon, G. Stewart, and K. C. Ikeler. 1928. *Sheep Ranching in Utah*. Utah Agricultural Experiment Station Bulletin No. 204. 60 pp.
Fulcher, G. and O. L. Brough, Jr. (Eds.) 1961. *Water and Range Resources and Economic Development of the West. Economic Analysis of Multiple Use: The Arizona Watershed Program—A Case Study of Multiple Use*. Western Agricultural Economics Research Council Report No. 9, Tucson, AZ. 141 pp. (mimeo.).
Goodsell, W. D. and J. R. Gray. 1961. *Cattle Ranches, Organization, Costs, and Returns*. Economic Research Service. USDA Agricultural Economics Report No. 1.
Gray, J. R. 1968. *Ranch Economics*. Iowa State Univ. Press, Ames, IA. 534 pp.
Gray, J. R. 1969. *Economic Research in the Use and Development of Range Resources—Range and Ranch Economics Bibliography*. Western Agricultural Economics Research Council Report No. 11. Las Cruces, NM. 199 pp. (mimeo.).
Gray, J. R. 1983. An historic look at range economics research. *Proceedings—Range Economics Symposium and Workshop. Aug 31–Sept. 2, 1982*. U.S. Forest Service General Technical Report No. INT-149. pp. 1–11.
Gray, J. R., T. M. Stubblefield, and K. N. Roberts. 1965. *Economic Aspects of Range Improvements in the Southwest*. New Mexico Agricultural Experiment Station Bulletin No. 498.
Heady, H. F. 1975. *Rangeland Management*. McGraw-Hill, New York. 460 pp.
Hockmuth, H. R. 1952. Economic aspects of range management. *J. Range Manage.* 5:62.
Hockmuth, H. R. and W. D. Goodsell. 1948. *Commercial Family Operated Cattle Ranches, Intermountain Region, 1930–47*. Bureau of Agricultural Economics, USDA Farm Management. 71 pp.
Hooper, J. F. 1969. Economics of fertilization and rates of grazing in California grassland management. Ph.D. dissertation, Univ. of California, Berkeley.
Hopkin, J. A. 1954. Economics of western range resource use. Ph.D. dissertation, Iowa State College, Ames.
Jameson, D. A., S. A. D'Anquino, and E. T. Bartlett. 1974. *Economics and Management Planning of Range Ecosystems*. 1st ed. Balkema, Rotterdam, The Netherlands. 244 pp.
Lansford, R. R. and J. R. Gray (Eds.). 1968. *Forces Restructuring Production and Marketing in Commercial Agriculture*. Western Agricultural Economics Research

Council Range Resources Committee Report No. 10. Tucson, AZ. 254 pp. (mimeo.).

Lloyd, R. D. 1959. Costs and returns from seeding publicly-owned sagebrush grass ranges to crested wheatgrass. Ph.D. dissertation, Utah State University, Logan.

McConnen, R. J. 1960. The economics of range fertilization in California. Ph.D. dissertation, Univ. of California, Berkeley.

McCorkle, C. O., Jr. (Ed.). 1959. *Economic Research in the Use and Development of Range Resources—Economics of Range and Multiple Use.* Western Agricultural Economics Research Council Report No. 2, Logan, UT. 159 pp. (mimeo.).

McCorkle, C. O., Jr. and D. D. Caton. 1962. *Economic Analysis of Range Improvement. A Guide for Western Ranchers.* California Agricultural Experiment Station, Giannini Foundation Research Report. 255 pp.

Morton, G. E. 1914. *Cost of Beef Production Under Semi-range Conditions.* Colorado Agricultural Experiment Station Bulletin No. 189. 6 pp.

Nielsen, D. B. 1965. Economics of federal range use and improvement for livestock production. Ph.D. dissertation, Oregon State University, Corvallis.

Nielsen, D. B. 1967. *Economics of Range Improvements—A Rancher's Handbook to Economic Decision Making.* Utah Agricultural Experiment Station Bulletin No. 466. 49 pp.

Nielsen, D. B. (Ed.) 1967. *Economic Research in the Use and Development of Range Resources—Range and Ranch Problems, Policy Implications, and Alternatives for Future Research.* Western Agricultural Economics Research Council Report No. 9, Reno, NV. 183 pp. (mimeo.).

Osgood, E. S. 1929. *The Day of the Cattleman.* Univ. of Minnesota Press, Minneapolis. 234 pp.

Pearse, C. K. and A. C. Hull. 1943. Some economic aspects of reseeding range lands. *J. Forestry* 41:346–358.

Roberts, N. K. (Ed.) 1962. *Economic Research in the Use and Development of Range Resources—Inter-use Competition for Western Range Resources.* Western Agricultural Economics Research Council Report No. 4, Reno, NV. 179 pp. (mimeo.).

Roberts, N. K. (Ed.) 1963. *Economic Research in the Use and Development of Range Resources—Development and Evolution of Research in Range Management Decision Making.* Western Agricultural Economics Research Council Report No. 5, Laramie, WY. 186 pp. (mimeo.).

Roberts, N. K. (Ed.) 1964. *Economic Research in the Use and Development of Range Resources—Measuring the Economic Value of Products from the Range Resource.* Western Agricultural Economic Research Council Report No. 6, Reno, NV. 145 pp. (mimeo.).

Roberts, N. K. and J. R. Gray (Eds.) 1961. *Economic Research in the Use and Development of Range Resources—Adjustments in the Range Livestock Industry.* Western Agricultural Economics Research Council Report No. 3, Fort Collins, CO. 152 pp. (mimeo.).

Samuelson, P. A. 1964. *Economics: An Introductory Analysis.* 6th ed. McGraw-Hill, New York. 838 pp.

Saunderson, M. H. 1950. *Western Stock Ranching.* Univ. of Minnesota Press, Minneapolis. 247 pp.

Smith, H. R. 1910. *Economical Beef Production.* Nebraska Agricultural Experiment Station Bulletin No. 116. 49 pp.

Stoddart, L. A. and A. D. Smith. 1943. *Range Management.* 1st ed. McGraw-Hill, New York.

Stoddart, L. A. and A. D. Smith. 1955. *Range Management*. 2nd ed. McGraw-Hill, New York.

Stoddart, L. A., A. D. Smith, and Thadis W. Box. 1975. *Range Management*. 3rd ed. McGraw-Hill, New York. 532 pp.

Subcommittee on Range Research Methods of the Agricultural Board. 1962. Economic Research in Range Management. In: *Basic Problems and Techniques in Range Research*. National Academy of Science Publication No. 890. Pp. 193–221.

Upchurch, M. L. 1953. Economic factors in western range improvements. *J. Farm Manage.* 35:728.

Vass, A. F. 1926. *Range and Ranch Studies in Wyoming*. Wyoming Agricultural Experiment Station Bulletin No. 147.

Wagstaff, F. J. 1983. Program Chairman's Foreword. *Proceedings—Range Economics Symposium and Workshop. Aug 31–Sept. 2, 1982*. U.S. Forest Service General Technical Report No. INT-149.

Wennergren, E. B. (Ed.) 1965. *Economic Research in the Use and Development of Range Resources—Goals and Public Decision-Making in Range Resource Use*. Western Agricultural Economics Research Council, Report No. 7, Portland, OR. 169 pp. (mimeo.).

Wennergren, E. B. (Ed.) 1966. *Economic Research in the Use and Development of Range Resources—1. Recreation Use of the Range Resources and 2. Decision Theory Models in Range Livestock Research*. Western Agricultural Economics Research Council Report No. 8, San Francisco, CA. 186 pp. (mimeo.).

Chapter 2

Economics in a Range Management Setting

IMPORTANCE OF ECONOMICS TO THE RANGE MANAGER

Range managers are sometimes initially suspicious of range economists—the manager of private rangeland largely because range economists necessarily ask questions involving personal information (costs, returns, production rates) and the public range manager because he often makes management decisions based on ecological principles and he is understandably not anxious to have such decisions tested from an economic viewpoint. Few range managers, public or private, want to hear that they could not afford some past management decision or even that they cannot afford some future course of action to which they are already committed.

Fulcher (1959) defined the goal of the public range manager as managing public range resources to achieve "...sustained [forage] yield over time," as contrasted with the private range manager who attempts to adjust forage yield and utilization in order to "...maximize income over time." In the same report, Caton (1959) more rigorously defined the goal of the private range manager as attempting "to maximize the present value of the flow of net ranch income over a foreseeable time period" subject to some "tolerance limit" involving the uncertainty of receiving a minimum level of income for servicing of a debt. Hopkin (1959) offered a more general and idealized "societal definition" of the goal of range management as "...maximum net product over the planning period of society" and pointed out that such a goal would not necessarily correspond to that of obtaining maximum profit over the planning horizon of the individual rancher.

If it can be generally agreed that, for economic survival of the ranch, the private manager must concern himself with maximum net ranch income over some definite time period (perhaps his lifetime or that of his children) and that the public range manager is usually charged to direct his efforts toward maximum sustained yield, it is obvious that both groups need the tools of economic analysis to achieve "efficient allocation of scarce resources between competing uses." Both private and public range managers work with limited or scarce resources (land, range improvement budget, vehicles, personnel) and devoting these scarce resources to one use necessarily precludes alternative uses. If harvested completely by yearling steers, spring forage on a private ranch operation cannot be used to also support brood cows any more than federal land agency capital improvement funds spent to construct access roads can be used again to develop stock water facilities. In recent decades range scientists have developed numerous improvement practices and range uses that are both biologically sound (they have been proven to work) and economically feasible (they have been shown to pay). However, no range manager has sufficient resources to implement all feasible projects or uses possible on rangeland under his control and he must choose from among those available. Economics, sometimes called the science of choice, offers a means of systematically evaluating and selecting from among all possible alternatives.

This trade-off between possible alternative uses of resources may often lead to painful decisions. For example, suppose a small rural town faces the decision of whether to spend limited tax proceeds to reduce air pollution from a coal-fired steam electrical generator or instead to devote the tax funds to purifying the city's culinary water supply. Despite the fact that the town's citizens value both clean air and clean water, the choice between them must be made.[1]

Because of the unavoidable trade-off relationships involved in the use of public rangeland funds, even decisions involving what might be called critical[2] projects fall within the realm of economic analysis. Suppose a public range manager has, on land under his control, a highly visible sore spot in the form of a steep area that has become denuded and badly eroded by indiscriminate off-road vehicle use. Even though it is agreed that the monetary returns from revegetating the area will be far less than the costs of soil preparation and seeding, it is also agreed by the land management agency (and the local public observing its actions) that the present situation is an intolerable one and that something must be done immediately to reclaim the area. Even in this situation, in which few people would disagree that revegetation must be done, public monies are to be used to fund the reseeding. Since monies used for this project become permanently unavailable for funding of alternative useful projects, the land manager is obligated to seek a "least-cost" means of accomplishing revegetation. What began as a biological problem has become an economic decision.

SETTING THE STAGE—WHY RANGELANDS ARE GRAZED

About 73 percent of all public land in the contiguous 11 western states is grazed by domestic livestock (Nielsen and Workman, 1971). The percentage of privately owned land in the 11 western states that is grazed is probably even higher. Why is this land grazed by livestock rather than used in some other way? Range forage is a "flow"[3] resource that with present technology can be converted on a large scale to products useful to man only through livestock grazing. Proper grazing during one season does not adversely affect the level of forage production during succeeding years and most rangelands of the United States and the world can be managed for sustained yield in perpetuity. It should be emphasized, however, that many rangelands are grazed not because they are well adapted to production of forage for harvest by domestic livestock, but because they are not well adapted to other land uses. The late Dr. L. A. Stoddart was careful to point out to students enrolled in his range management technical problems class at Utah State University that rangelands are grazed because they are too steep, too dry, too salty, too high, too rocky, or "too something" to be used for anything else and that they are often not well suited to grazing either. Thus many rangelands have become traditional grazing areas by default and have only recently been discovered by the general public and considered for other uses such as outdoor recreation. This relationship is explained by the law of comparative economic advantage, which states that a particular product will be produced in a given area because resources available in that area make it more advantageous for production of the product there than in any other area. Because of favorable soil and climate, for example, corn grows better in the Midwest than anywhere else in the United States. Two types of economic advantage may exist for any particular area, *absolute advantage* and *comparative advantage*. Thus, in addition to corn, the Midwest can also produce wheat better than can the western Great Plains states. The Midwest can also produce more cattle per acre than can the 11 western states. The Midwest, then, holds the absolute economic advantage for the production of all three crops. But why does the Midwest produce corn rather than wheat and cattle? The answer lies in the law of comparative economic advantage. While the Midwest can outproduce other areas in all three crops, it can outproduce them more in terms of net return per acre with corn than with wheat or cattle.

Gray (1968) discusses another version of this concept, the law of least *comparative disadvantage*, in explaining why range areas of the western states are used to produce livestock. While numerous other areas can outproduce the range states in almost any commodity mentioned, rangelands are at less of a comparative economic disadvantage with livestock than with other products.

Thus these generally unproductive lands are grazed by livestock simply because there is no productive alternative use. One possible exception is recreation, which at present actually competes with grazing on only a very small portion of the range area.

THE WESTERN CATTLE RANCH—A SKETCH OF THE OPERATION

In terms of land, western livestock ranches are large. Measured only in acres owned or grazed, even small western ranches appear huge in comparison with other agricultural operations in the United States. Perhaps the most striking feature of western ranches, however, is their nonhomogeneity; no two are alike. In some regions, particularly the Great Plains, ranching operations are largely confined to privately owned cropland and rangeland with perhaps some leased private land or "school sections" leased from the state. More common along the eastern foothills of the Rockies, in the Southwest, throughout the intermountain area, and in the western foothills of the Sierra-Nevada are operations involving a mixture of private land as a base, supplemented with federal grazing leases (Bureau of Land Management spring or fall foothill range or desert winter range and U.S. Forest Service mountain summer or spring–fall foothill range) along with state leases.

Types of operation are numerous and varied throughout the West. They depend upon topography, climate, type of vegetation, transportation facilities, distance to market, and preferences and special skills of the manager. Included are pure bred cattle, commercial cow–calf, cow–yearling, stocker cattle, pure bred sheep, self-contained (plains) ewe–lamb, and migratory (intermountain) ewe–lamb operations. Large capital investments are required by these ranches. The 300-unit breeding herd (300 brood cows or 1500 breeding ewes), generally considered a minimum sized self-sufficient family operation (Gray, 1968), has recently required an investment of around three-quarters of a million dollars in land, livestock, improvements, and machinery (Capps and Workman, 1982; Workman and King, 1982) if a ranch investor were starting from scratch.[4] Rates of return on investment[5] in both cattle and sheep ranches have traditionally been low by nonranching standards except, perhaps, during the short period in the 1880s when rates of return varied between 25 and 40 percent (Osgood, 1929). Returns in recent years have commonly been in the neighborhood of 2 percent (Workman, 1981). Of course when real estate appreciation is included in the annual return calculations, the percent return on ranch investment has typically been a much more competitive 10–15 percent (Bostwick, 1967; McConnen, 1976; Workman, 1981).

A hypothetical but representative intermountain cow–calf ranch will next be described in detail. The description will include ranch operation procedures, land, livestock, machinery, and improvement inventory and value, production rates, returns, costs, and mortgage information. Hypothetical data presented are from 1980 and do not reflect current price or cost levels nor even current production rates. No general conclusions concerning the current economic status of intermountain cattle ranches are to be drawn. Instead the description and analysis of a typical family sized Utah cattle ranch in 1980 are presented simply as a means of providing a guide to procedures commonly used to perform economic analyses of range livestock firms. Substitution of specific investment, production, cost, and price data into this guide will allow the reader to perform his own economic analysis of individual ranches in his own area.

Amounts of investment by category for a typical 300 cow Utah cattle ranch are shown in Table 2.1. It will be noted that the largest investment is in land (range, seeded pasture, irrigated cropland) followed by livestock, improvements, federal grazing permits [Bureau of Land Management (BLM) and U.S. Forest Service (USFS)] and machinery, respectively. Required total investment exceeds three-quarters of a million dollars or nearly $2800 per cow unit.

Our representative 300 cow intermountain ranch runs 1 bull for each 20 cows, or a total of 15 bulls, which are retained in the herd for three years and then replaced. Cows are bred in June, July, or August and calves are born March–May. Calves are weaned about mid-October. Except for heifer calves kept for cow herd replacements, calves are sold about November 1 at an average age of 7 months. An average of 15 percent of the cow herd is culled annually and the required 45 replacement heifers are bred at about 14 months of age, giving birth to their first calf at an average age of 23 months. This breeding schedule allows the first calf heifers one extra estrus cycle to breed back, helping to insure that they stay on schedule with the rest of the cow herd. Calf crop percentage, expressed as the number of calves weaned in November divided by the number of breeding cows in the herd the preceding January 1, averages 80 percent.[6] Thus 240 calves, about one-half steers and one-half heifers, are weaned November 1 at average weights of 400 pounds for steers and 380 pounds for heifers. Resulting annual livestock production and sales

Table 2.1 Total and Per Cow Investment by Category for a Hypothetical 300 Cow Utah Cattle Ranch, 1980

Category of investment	Total investment ($)	Per cow investment ($)
Land	482,000	1607
Federal grazing permits	58,000	193
Buildings and improvements	60,000	200
Machinery and equipment	50,000	167
Livestock		
Cattle	178,500	595
Horses	1,500	5
Total	830,000	2767

Source: Workman (1981).

Table 2.2 Annual Production and Revenue for a Hypothetical 300 Cow Utah Cattle Ranch, 1980

Number and category	Sale weight (lb)	Sale price ($/lb)	Total revenue ($)	Revenue per cow ($)
45 cull cows	1000	0.35	15,750	52.50
5 cull bulls	1500	0.40	3,000	10.00
75 heifer calves	380	0.54	15,390	51.30
120 steer calves	400	0.58	27,840	92.80
Total			61,980	206.60

Source: Workman (1981).

are shown in Table 2.2. Death loss was ignored to facilitate reader ease of following calculations of calf crop percent and cow and bull replacement rates. The $61,980 of total annual revenue represents the easy half of the ranch income statement. We now turn to the more difficult half, that of calculating annual costs.

THE STANDARD RANCH INCOME STATEMENT

A standard 12-month ranch income statement is displayed in Table 2.3. Gray (1968) described the income statement as a "moving picture" of ranch transactions and considered it to be the most useful and revealing short-term record that can be developed by the rancher.

Total *annual cash returns*, $70,000, consist of all receipts from livestock or crops sold. Subtracting *annual cash costs*, $31,000 (all cash operating expenses except loan interest and principal payments), yields *net cash ranch income*. It is this amount of cash that is available to purchase new machinery and improvements, provide for family living expenses, and to pay principal and interest on any outstanding loans against land, improvements, livestock, and machinery.

Next *depreciation costs* are subtracted to form *net ranch income*. Depreciation is the gradual but inevitable wearing out of all improvements and equipment, no matter how well they are maintained. Depreciation is often called a noncash cost meaning that it does not have to be paid in cash each year (and can occasionally be postponed for several years in a row). Still, the cost of replacing fences, trucks, etc., that are finally completely worn out must eventually be paid. The calculation of depreciation is simply an accounting technique to systematically convert large future expenditures into more easily managed annual costs. During years of unfavorable prices or drought there is a natural tendency to postpone setting funds aside for replacement of improvements and machinery. This common practice is sometimes termed "living on depreciation." At best, however, this is only a stalling procedure. Failure to reserve the necessary funds only makes future replacement more difficult.

Table 2.3 Standard Income Statement for a Hypothetical 300 Cow Utah Cattle Ranch, 1980

	Total ($)	Per cow ($)	
Annual cash returns			
Cattle sold	61,980	206.60	
Crops sold	8,020	26.73	
Total	+70,000		233.33
Annual cash costs			
Feed, including grazing fees	3,500		
Labor hired	6,000		
Machinery repairs	2,000		
Building and improvement repairs	1,000		
Veterinary	500		
Taxes	5,000		
Crop expense	6,000		
Bull purchase	5,000		
Other	2,000		
Total	−31,000		103.33
Net cash ranch income	39,000		130.00
Depreciation costs			
Building and improvements depreciation	2,000		
Machinery and equipment depreciation	5,000		
Total	−7,000		23.33
Net ranch income	32,000		106.67
Allocation of Net Ranch Income to Labor			
Interest on capital			
Real estate (10%)	60,000		
Livestock and equipment (10%)	23,000		
Total	−83,000		
Net return (loss) to operator and family labor	(51,000)		
Allocation of Net Ranch Income to Capital			
Value of operator and family labor	−15,000		
Net return to investment	17,000		
Percent return on $830,000 total investment		2.05	

Source: Workman (1981).

Several quite complex methods of calculating depreciation are currently in widespread use, particularly for real estate appraisal and federal income tax purposes. However, for the purpose of constructing a ranch income statement, the simple straight-line method is sufficiently accurate. This method consists of subtracting estimated salvage value of a machine or building from original cost (or current value) and dividing by expected useful life. The result is the portion of original cost or value of the asset that is exhausted during an average year of its productive life. For example, annual depreciation for a new $11,000 truck with $1000 salvage value at the end of its 5 year expected life is ($11,000 − $1000) : 5 = $2000.

Gray (1968) recommends an alternative depreciation accounting technique. He includes the entire expenditure for machinery or construction of improvements during a particular year in annual cash costs. He then recognizes the fact that only a portion of the total investment expenditure is attributable to the current year by inclusion of another category in the ranch income statement called the net change in inventory. As implied by the name, this adjustment to costs and returns expresses the combined effects of increased inventory value due to purchases of new machinery and improvements (or substantial improvements to existing assets which add to their market value) and decreased inventory value due to asset depreciation. Thus purchase of the $11,000 truck mentioned above would reduce net ranch income by $2000 using either Gray's approach or the method shown in Table 2.3. With Gray's method annual cash costs are increased by the $11,000 purchase price of the truck, while the net change in inventory is an increase of $9000 ($11,000 original value minus $2000 first year depreciation) yielding a $2000 decrease to net ranch income. With the method of Table 2.3, the $11,000 purchase is not shown as an annual cash cost and the only change brought about by the new truck is a $2000 increase in depreciation which also reduces net ranch income by $2000.

Subtracting $7000 total depreciation costs yields a net ranch income of $32,000.[7] These are the returns available for mortgage payments and family living expenses. Standard income statements next allocate net ranch income to its two claimants, labor and capital. In Table 2.3, allocation is first made to labor by subtracting *interest on capital* (10 percent of the value of land, improvements, equipment, and livestock) and specifying the remainder, a negative $51,000, as compensation for operator and family labor. Allocation to labor is based on the premise that all capital (including the owned or equity portion of real estate and personal property) must be paid a fair return on investment.

The second allocation of net ranch income in the standard income statement is to capital. This is accomplished by subtracting the *value of operator and family labor*[8] (3750 hours at $4 per hour in this example) from net ranch income and attributing the $17,000 remainder as a return to total ranch investment. Dividing this remainder by the $830,000 total value of land, improvements, livestock, and equipment gives a rate of return on total ranch investment of 2.05 percent. The premise underlying allocation to capital is that operator and labor must be compensated at market rates. In reality, both compensation for operator and family labor (when allocating the residual to capital) and interest on owned capital (when allocating the residual to labor) are opportunity costs. Thus neither operator labor nor owned capital has to be fully paid, provided, of course, there is sufficient net ranch income for the family to live on.

Although it provides extremely interesting information, the standard income statement does not answer several important and frequently asked questions. It offers no explanation of how today's ranchers stay in business while receiving extremely low returns on large investments in land and improvements. Nor does the standard income statement furnish an explanation as to why anyone would invest in ranch property in the face of such a bleak cost–price picture. Analyzing and understanding the financial aspects of ranching operations requires answers to two simple but important questions. First, will the ranch produce sufficient net income for the ranch family to live on after all operating expenses (including loan service) have been paid? Second, how much net ranch income (including real estate appreciation) is available to compensate owned capital (equity)? Since neither of these crucial questions is answered by the standard ranch income statement of Table 2.3, I propose the use of the modified income statement below.

THE MODIFIED RANCH INCOME STATEMENT

As in the standard income statement, formulation of the modified income statement begins with the calculation of net ranch income by subtracting annual cash and depreciation costs from annual cash returns (Table 2.4). Next an answer is provided to the first question posed above: "Will the ranch produce enough net income for the family to live on?" The answer comes from explicit recognition of *loan service costs* that, when subtracted from net ranch income, yield *net return available for family living expenses*. The $5909 combined principal and interest payment for real estate is based on a 6 percent 30-year loan established 20 years ago (in 1960). The $116,190 original purchase price was calculated by adjusting the current real estate value ($600,000) for the 8.55 percent average annual appreciation that has occurred in rangeland prices during the 20-year period 1960–1980 (USDA, 1980). Down payment was set at 30 percent of purchase price ($34,857), leaving $81,333 as the initial principal balance. Using the Present Worth of One per period annuity factor from Table 8.5 (Chapter 8), annual loan payment is $81,333/13.765 = $5909. Loan service costs of $13,715 for working capital (livestock and equipment) are based on a 9 percent 10-year loan established in 1975. The $125,750 original purchase price was calculated by adjusting the current working capital value ($230,000) back five years for livestock and equipment price increases during the 1975–1980 period. Down payment was set at 30 percent of purchase price ($37,725), leaving $88,025 as the original principal balance. Referring again to

ECONOMICS IN A RANGE MANAGEMENT SETTING

Table 2.4 Modified Income Statement for a Hypothetical 300 Cow Utah Cattle Ranch, 1980

Item	Dollars
Annual cash returns	+70,000
Annual cash costs	−31,000
Depreciation costs	− 7,000
Net ranch income	32,000
Loan service costs	
Real estate	5,909
Working capital	13,715
Total	−19,624
Net return available for family living expenses	12,376
Land appreciation	+51,300
Payment to mortgage principal	+11,290
Gross proceeds to ranch investment	74,966
Value of operator and family labor	−15,000
Net proceeds to owned ranch capital	59,966
Percent return on $733,164 owned ranch capital	8.18%

Source: Workman (1981).

Table 8.5, annual loan payment is $88,025/6.418 = $13,715. Subtracting total loan service cost ($19,624) from net ranch income yields $12,376 as the net return available for family living expenses. Combined with utility and automobile expenses already included in annual cash costs along with the perquisites mentioned above it appears that this amount was sufficient to cover family living costs in 1980. This measure of net cash returns might be considered short-term annual net income.

To convert this short-term income to long-term annual net income, the modified income statement introduces two additional income sources. First, *land appreciation* is included, calculated in our example at the 8.55 percent average annual compound rate for rangeland mentioned above. The purpose of including the $51,300 increase in real estate value (8.55 percent × $600,000) as income is not to encourage inflation but to realistically acknowledge that inflation is occurring. Current land prices, apparently much higher than those justified by crop or livestock production, are perhaps the best evidence that expected land appreciation is at least as important to the investor as crop or livestock income (Workman and King, 1982).

The second additional income source included in the modified income statement is *payment to mortgage principal*. This inclusion is based on the recognition that a principal payment is actually a payment from the borrower to himself since $1 paid toward principal increases the borrower's equity (the difference between the market value of an asset and the amount owed on it) by $1. The $11,290 contribution to principal in Table 2.4 is the sum of the $3113 principal payment for year 20 of the 30-year real estate loan and $8177 for year 5 of the 10-year livestock and equipment loan. Like land appreciation, principal payment returns are not available to the borrower until he either refinances or sells the ranch property. Still, such gains in equity are just as important to the ranch investor as cash income.

The modified income statement includes one final item to complete the analysis. As discussed for the standard ranch income statement above, the contribution that labor makes to *gross proceeds to ranch investment* must be identified. Thus to calculate the income attributable solely to *ownership* of land and livestock, income rightfully belonging to labor must be subtracted. Of course, if the ranch owner and operator were different people, the $15,000 value of operator and family labor would be included in annual cash costs and this last step would not be necessary. *Net proceeds to owned ranch capital*, then, total $59,966 annually: We have an answer to the second question posed above. This $59,996 represents the long-term annual net income available to compensate the $733,164 of owned capital [this owned capital, or equity, is the sum of the market value of real estate ($600,000) minus the year 20 real estate loan balance, ($43,479) and the market value of livestock and equipment ($230,000) minus the year 5 working capital loan balance ($53,357)]. The resulting 8.18 percent rate of return on owned capital accounts for all income (both cash and equity growth) and all costs (both cash and opportunity). This rate of return can be compared directly with nominal rates of return on other long-term investments.[9]

The modified ranch income state is recommended over the standard income statement since it is superior to it in several important ways: (1) It is easier to interpret—net return is expressed in terms of net income actually available to live on. (2) It is more realistic—real estate appreciation is included as income. From the investor's viewpoint such income is just as important as that derived from the sales of calves. (3) It is much simpler—hypothetical required rates of return on capital are not included. Instead the borrower or lender can compare the rate of return actually earned on owned capital with his own concept of a "required" rate. The modified income statement of Table 2.4 will be referred to again in the range improvement analysis of Chapter 9.

THE RANCH FINANCIAL STATEMENT

The ranch financial statement or balance sheet has been called a "snapshot" of the financial status of the ranch as opposed to the "moving picture" character of the ranch income statement (Gray, 1968). The financial statement portrays the financial health of the ranch at a specific point in time.

For purposes of analyzing the economic effects of range improvements, changes in management practices, or expansion of the ranch operation through purchase of additional land, the financial statement provides no information not already found in the ranch income statement discussed above. However, a current financial statement is probably the record most often requested by lending institutions as part of the rancher's loan application. Thus it behooves anyone concerned with implementing range improvements and advanced management practices to become skilled in the preparation of the ranch financial statement. This record is often the key to obtaining the borrowed capital necessary to convert a sound idea to a sound on-the-ground practice.

A financial statement for the hypothetical Utah cattle ranch described above is shown in Table 2.5. Based on degree of liquidity, assets are placed in three categories: current, working, or fixed. Cash held in checking and savings accounts and the current market value of all other assets that will be or could be converted to cash in the immediate future are listed as current assets. In addition to the items shown, off-ranch investments such as stocks and bonds would also be included. The current values of all items in the livestock and machinery inventory appear as working assets. Fixed assets include the estimated market value of real estate, including land, improvements, and federal grazing permits. Although there may be some doubts about the legitimacy of the latter item, it should be included as a tangible asset for at least two reasons. Pragmatically, lending agencies do recognize grazing permits as security for real estate loans. The tangible character of grazing permits is also demonstrated by the fact that these assets are bought and sold (transferred with dependent livestock or dependent "base property") and do assume established market values as part of the price for dependent livestock or private land transferred along with the grazing permits.

The liabilities side of the ledger is also composed of current, working, and fixed categories, based upon how soon the debts are due. Current liabilities are those now due, such as feed and gasoline billed against a charge account, and those due in the immediate future, such as an operating loan made in midsummer and due upon fall sale of calves. Working liabilities are debts against assets with productive lives of several years. Such debts are normally retired over a period of time corresponding roughly to the productive life of the asset. Outstanding mortgages against long-lived or fixed real estate assets are listed as fixed liabilities. The three categories are then totaled and *net worth* is included as a liability simply to make total liabilities equal total assets. Net worth can be thought of as "debt-free" or owned assets invested in or loaned to the ranch (Gray, 1968). Thus net worth is a debt like any other asset borrowed by the ranch and should logically be listed as a liability.

It is advisable that the rancher prepare an accurate financial statement with the current date and his notorized signature prior to meeting with his banker to apply for a loan. Accuracy and completeness of the financial statement are crucial. This is the primary document supporting the rancher's

Table 2.5 Financial Statement for a Hypothetical 300 Cow Utah Cattle Ranch, December 31, 1980

Assets	
Current	
Cash	$ 1,000
Accounts receivable	1,000
Hay and grain	2,000
Market livestock	4,000
Working	
Breeding livestock	178,500
Horses	1,500
Machinery and equipment	50,000
Fixed	
Buildings and improvements	60,000
Federal grazing permits	58,000
Land	482,000
Total assets	$838,000
Liabilities	
Current	
Accounts payable	3,000
Operating loan	5,000
Working	
Livestock loan	39,779
Machinery loan	13,578
Fixed	
Real estate loan	43,479
Total	$104,836
Net worth	733,164
Total liabilities	$838,000

Source: Workman (1981).

request for required range improvement and management funds, and it forms the basis for continued mutual trust between the rancher and the lending institution.

LITERATURE CITED

Bostwick, D. 1967. Financial returns in agriculture. *Am. J. Agr. Econ.* 51:662–664.

Capps, T. L. and J. P. Workman. 1982. *Management, Productivity, and Economic Profiles of Two Sizes of Utah Cattle Ranches.* Utah Agricultural Experiment Station Research Report No. 69. 20 pp.

Caton, D. D. 1959. Selection of optimum season and intensity of grazing. In: *Economic*

Research in the Use and Development of Range Resources. Economics of Range and Multiple Use. Western Agricultural Economics Research Council Report No. 2, Davis, CA. Pp. 41–59.

Fulcher, G. D. 1959. Establishing range input–output relationships for economic analysis. Discussion of paper. In: *Economic Research in the Use and Development of Range Resources. Economics of Range and Multiple Use*. Western Agricultural Economics Research Council Report No. 2, Davis, CA. Pp. 33–39.

Gray, J. R. 1968. *Ranch Economics*. Iowa State Univ. Press, Ames. 534 pp.

Hopkin, J. A. 1959. Economic versus ecological criteria for deciding season and intensity of grazing—a discussion. In: *Economic Research in the Use and Development of Range Resources. Economics of Range and Multiple Use*. Western Agricultural Economics Research Council Report No. 2, Davis, CA. Pp. 73–77.

McConnen, R. J. 1976. *Public Land Grazing and Ranch Economics*. Agricultural Economics and Economics Department Staff Paper 76-10, Montana State University, Bozeman. 40 pp.

Nielsen, D. B. and J. P. Workman, 1971. *The Importance of Renewable Grazing Resources on Federal Lands in the 11 Western States*. Utah Agricultural Experiment Station Circular No. 155. 44 pp.

Osgood, E. S. 1929. *The Day of the Cattleman*. Univ. of Minnesota Press, Minneapolis. 234 pp.

U.S. Department of Agriculture. 1980. *Farm Real Estate Market Developments*. USDA Economics, Statistics, and Cooperatives Service Report No. CD-85, August. Washington, D.C. 46 pp.

Workman, J. P. and J. F. Hooper. 1971. Cost-size relationships of Utah cattle ranches. *J. Range Manage.* 24:462–465.

Workman, J. P. 1981. Analyzing ranch income statements—a modified approach. *Rangelands* 3:146–148.

Workman, J. P. and K. H. King. 1982. Utah cattle ranch prices. *Utah Sci.* 43:78–81.

TEXT NOTES

1 It is not difficult to also imagine a biological trade-off in such a situation in addition to the described economic trade-off. Perhaps the air pollution problem could be solved by the installation of a smokestack scrubber. However, at best such an installation uses water that otherwise could be made available to the town's citizens by processing it through the culinary system and, at worst, the spent water from the smokestack scrubber finds its way into the culinary water source and again reduces the available quantity of clean water.

2 Critical projects are defined as those that are considered necessary for some noneconomic reason despite the fact that they may not be feasible from a strictly economic point of view.

3 Flow or renewable resources are those for which different units become available for use during different time periods. In contrast, stock or nonrenewable resources, such as coal, do not increase significantly in quantity with time. Flow resources not used during the time period in which they become available cannot be held indefinitely for future use. Forage left at the end of the grazing season in excess of that required to perpetuate the stand of vegetation is essentially wasted—its potential as forage dies with the season (Nielsen and Workman, 1971).

4   It should be recognized that many genuine family ranching operations have spanned several generations. Thus the $2500 per brood cow unit implied here actually represents current marked value of land, livestock, etc., rather than the amount actually invested in these assets at the time of purchase.
5   Rate of return on investment is calculated as total annual returns minus annual cash costs, depreciation of machinery and improvements, and operator and family labor, all expressed as a fraction of total investment.
6   This method of calculation places an upward bias on calf crop percentage since any "open" females detected by pregnancy evaluation in October and November are usually sold prior to taking the January 1 cow herd inventory. A more accurate, but seldom used calculation, expresses the number weaned as a percentage of the total number of cows exposed to bulls during the breeding season of the *preceding* year. Thus in our example, the 240 calves weaned would represent a calf crop of 70 percent (240 calves divided by a total of 345 breeding cows exposed made up of 300 mature cows and 45 yearling replacement heifers) rather than the 80 percent stated above. Although not as accurate, the January inventory approach to calf crop percentage has become the standard of the range livestock industry. Whatever the technique used, the details of the calculation should be specified to avoid misinterpretation (Gray, 1968).
7   Another item not shown in Table 2.3 but often added to net cash ranch income in calculating net ranch income is the value of perquisites. Included are the rental value of the ranch home and the market value of home grown meat, milk, eggs, and garden produce.
8   This represents the wages foregone by the ranch family working on their own ranch instead of performing the same duties for pay on someone else's ranch. Alternatively, these are the wages that would have to be paid hired management and labor to perform the duties of the ranch family.
9   The calculations underlying this rate of return include loan service, based on nominal mortgage rates, and expected annual land appreciation, also a nominal measure.

Chapter 3

Supply and Demand— The Basic Tools

The principles of supply and demand are undoubtedly the most important concepts in economics. Unfortunately, they are also some of the economic principles least understood by noneconomists and are often applied incorrectly. Supply and demand are invaluable aids in understanding what is currently happening, in planning how to improve the current situation, and in predicting what is likely to happen in the future.

Because of widespread use of the terms supply and demand in everyday conversation as well as in the popular media, most people think they know what is involved in these concepts whether or not they actually do. This chapter attempts to carefully define these terms, explain the underlying concepts, and provide examples of how they are useful in range management.

DEMAND

Demand is defined as the quantity of product willingly bought per unit of time at a specific price. Enlarging upon this basic concept, a *demand schedule* (which may be represented by a numerical table, graphical curve, or mathematical function) is defined as the various quantities willingly purchased per time period at each of the various possible product prices. The demand schedule is based on the *law of demand*, which states that as the price of a product or service increases, fewer units are purchased and vice versa. Table 3.1 illustrates this familiar law with hypothetical beef purchase data. At a price of $1.75 per pound, only 6.25 billion pounds of beef are purchased annually by consumers. If beef price drops to $1 per pound, annual purchases increase to 25 billion pounds. The inverse relationship between beef price and quantity purchased is based on the concept of "decreasing marginal utility." In the jargon of economics, the term *marginal* means added or successive. Utility, as used here, means usefulness or "want satisfying power" of a good or service. The concept of decreasing marginal utility is based on the idea that each successive unit of a good or service consumed by the purchaser adds less to his total satisfaction than did the previous unit. This phenomenon is readily apparent no matter what the good or service in question: pieces of chocolate cake, number of rock concerts, or pounds of beef. Thus the individual consumer will increase his beef purchases per time period only if beef price goes down. As a rational buyer, he is willing to trade cash for beef only if he perceives the quantity of satisfaction resulting from the last pound of beef purchased to be at least equal to the price per pound. Of course, tastes and income vary among the various buyers comprising the beef market. At the upper price of $1.75 per pound some consumers buy no beef at all because they prefer substitutes such as pork, fish, or poultry. Others prefer beef but, because of income constraints, buy less expensive sources of protein. Still other consumers have somewhat higher incomes and, preferring beef over other meats, they buy small quantities even at the upper price of $1.75. If the beef price happens to decrease to $1 per pound, there are three immediate simultaneous effects. First, consumers who have been buying some beef at the higher price now buy more. Second, those individuals who could not afford beef at the higher price begin to exercise their preference. Third, even those people whose preference is for other types of meat begin to substitute the now more competitively priced beef.

The information in Table 3.1 represents the aggregate purchases of all consumers in the nation corresponding to each beef price shown. This information can also be portrayed graphically as in Figure 3.1. The resulting hypothe-

Table 3.1 Hypothetical Demand Schedule for Beef

Beef price ($ / lb)	Quantity purchased (billion lb beef per year)
1.75	6.25
1.50	12.50
1.25	18.75
1.00	25.00
0.75	31.25
0.50	37.50
0.25	43.75

SUPPLY AND DEMAND — THE BASIC TOOLS

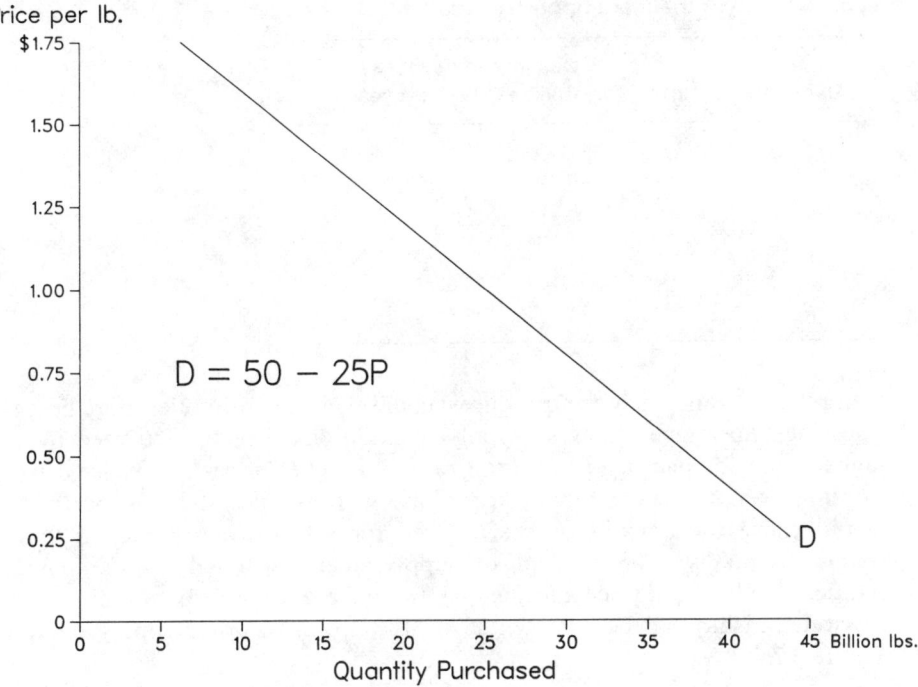

Figure 3.1 Hypothetical demand curve for beef.

tical demand curve[1] slopes from northwest to southeast, demonstrating why the law of demand is often referred to as the *law of downward sloping demand*.

SUPPLY

The concept of supply is analogous to the demand schedule of consumers, except that supply is based on the producer's wishes. Supply is defined as the quantity of a product willingly offered for sale per unit of time at a specific price. As with demand, the concept of supply is often represented as a *supply schedule* in the form of a numerical table, graphical curve, or mathematical function. A supply schedule may be defined as the various quantities willingly offered for sale per time period at each of the various possible product prices. The supply schedule is based on the concept of *increasing marginal costs*, which in economic jargon means simply that each additional unit of a product costs more to produce than did the preceding unit. This concept, which is explained more fully in Chapter 5, is illustrated in Table 3.2. At the lower packaged beef price of only $0.25 per pound, beef retailers are willing to offer only 1.56 billion pounds of beef for sale annually. To attempt to sell more than this amount would cost more in handling costs and payments to beef wholesalers, packers, feeders, and producers than would be earned in sales at the $0.25 price. If the beef price happened to increase to $1.00 per pound, each

Table 3.2 Hypothetical Supply Schedule for Beef

Beef price ($ / lb)	Quantity offered for sale (billion lb beef per year)
0.25	1.56
0.50	9.38
0.75	17.19
1.00	25.00
1.25	32.81
1.50	40.63
1.75	48.44

beef retailer[2] (and each beef producer) could then afford to offer more beef for sale since the higher costs of supplying additional beef are covered by the higher price for packaged beef (and resulting higher prices for feeder calves). For the beef market in total, the increase in beef price to $1.00 per pound would cause the amount of beef offered for sale to increase to 25 billion pounds annually. The concept of supply, then, is based on the positive relationship between price and the quantity offered for sale. In graphical form, this relationship can be represented by the upward sloping supply curve of Figure 3.2.[3]

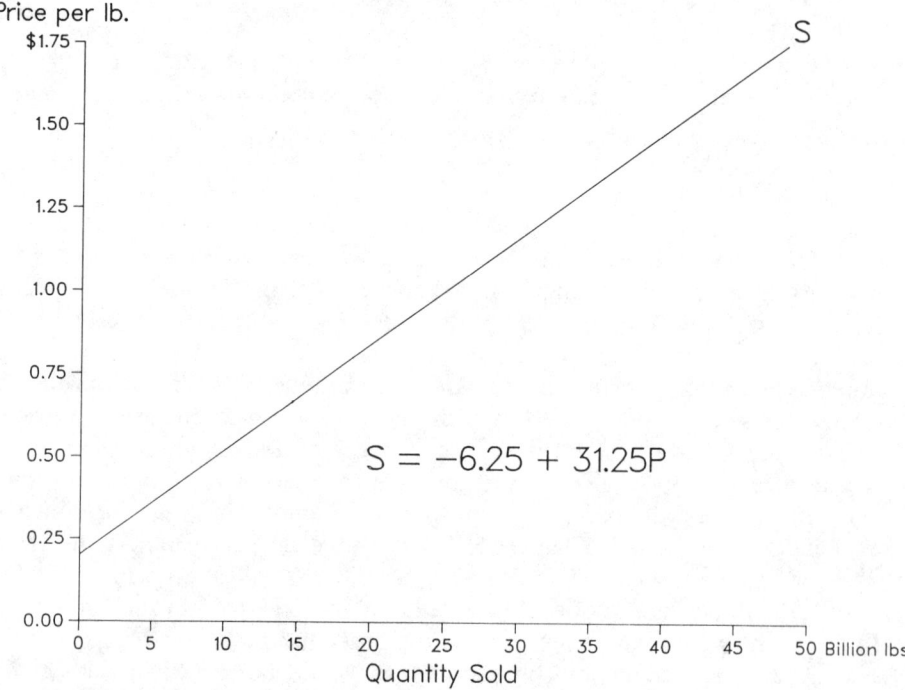

Figure 3.2 Hypothetical supply curve for beef.

SUPPLY AND DEMAND — THE BASIC TOOLS

MARKET EQUILIBRIUM

The concept of market equilibrium is often confusing when encountered for the first time. Equilibrium in the market for packaged beef is illustrated in Figure 3.3. The intersection of the supply curve, representing willingness to sell, with the demand curve, based on willingness to buy, gives an *equilibrium beef price* of $1.00 per pound and an *equilibrium quantity* exchanged of 25 billion pounds annually.[4] The market is at equilibrium in the sense that only at this price and at this quantity of exchange is the amount willingly offered for sale equal to the amount willingly bought. No matter what the price, the quantity actually sold will obviously be equal to the quantity actually bought. But only at the equilibrium price of $1.00 per pound is the amount consumers want to buy at the prevailing price the same as the amount producers want to sell at that same price. To illustrate the important difference between the quantity actually bought (and sold) and the willingness to buy and sell, suppose the market price is temporarily above the equilibrium at $1.25. At this price, producers are willing (and able, since they can now cover the higher unit costs associated with increased production) to sell 32.81 billion pounds per year. But at this higher price, consumers are willing to buy only 18.75 billion pounds annually. Of course producers can sell only as much as consumers are

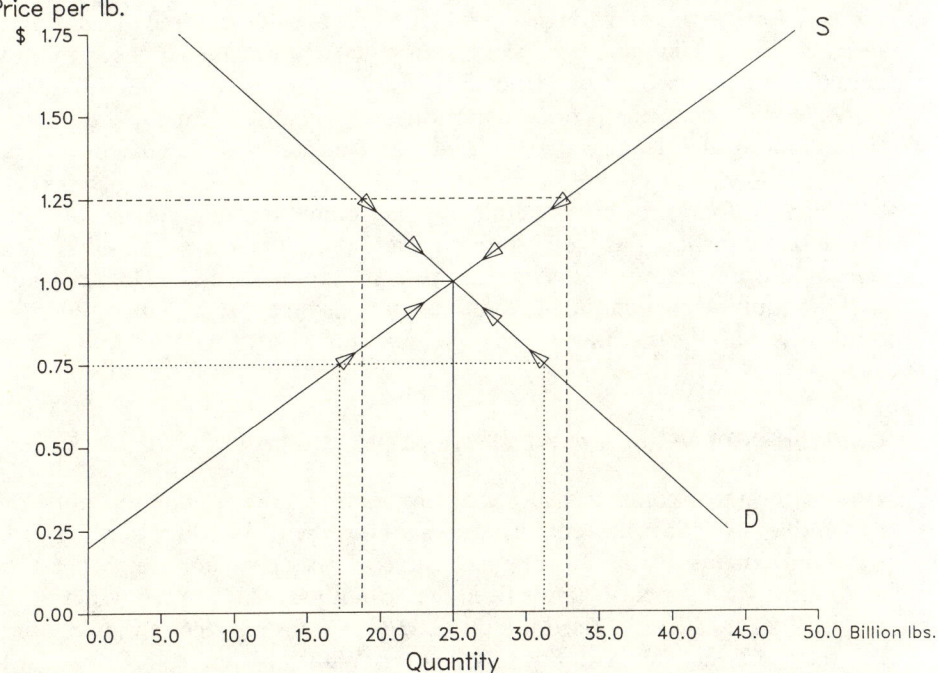

Figure 3.3 Hypothetical demand and supply curves for beef, equilibrium price, and equilibrium quantity.

willing to buy, 18.75 billion pounds, resulting in a surplus supply of 14.06 billion pounds. But how will beef producers (retailers, wholesalers, and packagers in this case) respond to this unpleasant situation? Faced with the possibility of spoilage, they will offer a portion of their surplus perishable product at a price lower than $1.25. And how will consumers respond to the lowered price? They will buy more, in effect moving downward and to the right along their demand curve. If the beef surplus does not clear the market entirely, there will be additional rounds of price lowerings by beef retailers followed by increased consumer purchases until actual price if forced down along the demand curve to again equal equilibrium price, simultaneously exhausting the surplus beef and forcing actual beef quantity supplied down and to the left along the supply curve to equal equilibrium quantity.[5]

Now suppose the market price was temporarily at $0.75 per pound, below the equilibrium price. At this lower price consumers are willing to buy 31.25 billion pounds but producers are willing to sell only 17.19 billion (since production costs must be covered by sales revenues, producers are also able to sell only 17.19 billion). Thus there is an *excess demand* of 14.06 billion pounds. How will consumers and producers respond to this market shortage? Individual consumers, disappointed that they cannot buy all the beef they want to buy at the low price will offer a slightly higher price to obtain a portion of the limited beef supply. Producers will interpret this slight increase as a signal to produce more beef, which they will be able as well as willing to do since the increased price will allow higher production costs to be covered. This process of consumer bidding and producer response will continue until once again the actual price has been pushed up along the supply curve to the equilibrium price. Since the increased price also pushes the quantity demanded up and to the left along the demand curve, the supply shortage is simultaneously alleviated.

Tremendous pressures operate in the competitive market to force the actual price to equal the equilibrium price and the actual quantity sold to equal the equilibrium quantity. It is only at this equilibrium price of $1.00 per pound and the equilibrium quantity of 25 billion pounds that everyone involved in the market is happy and there is no pressure for either price or quantity to change.[6]

CURVE SHIFTS VERSUS MOVEMENTS ALONG CURVES

One of the most common sources of problems in understanding supply and demand analysis involves confusion of shifts of the demand (or supply) curve with movements along the original curve. These problems might best be introduced by a simple example from the distant past. An English writer of the 1600s, Gregory King, noticed that in good years farmers received lower prices for their products than during poor years. While this effect on price is so obvious that even second grade school children are aware of it (Samuelson, 1964), it might be helpful to illustrate this phenomenon using supply and

SUPPLY AND DEMAND — THE BASIC TOOLS

demand curves. Suppose we use the amount harvested and the price of wheat as our example. When using supply and demand curves to predict or better understand the direction of change in price, the *initial equilibrium* situation should first be represented by an intersection of a supply curve with a demand curve. As shown in Figure 3.4, the supply curve for the poor year, S_P, intersects the demand curve for the poor year, D_P, at a price of $2 per bushel. Next the question should be asked: "Which curve, if either, will be changed by the changed market situation?" Will the wheat supply curve, representing willingness (and ability) to sell at each price, be different in a good year than in a poor year? Of course it will! Due to the cooperation of the weather, wheat producers are willing (and able) to sell more wheat at each price in a good year than in a poor year. Thus the supply curve shifts out in the good year from S_P to S_G. How about the demand curve? Will the fact that the weather is better this year for growing wheat than it was last year change the demand curve representing the willingness of consumers to buy at each price? Absolutely not![7] Since willingness to buy wheat is not changed by the favorable weather, the demand curve for wheat is the same in the good year as in the poor year. Thus the outward shift of the supply curve corresponds to a downward movement along the original demand curve. The price falls from $2 to $1.50 per bushel and, of course, more wheat is purchased at the lower price. But the main point is this: The quantities that consumers want to buy at either the old

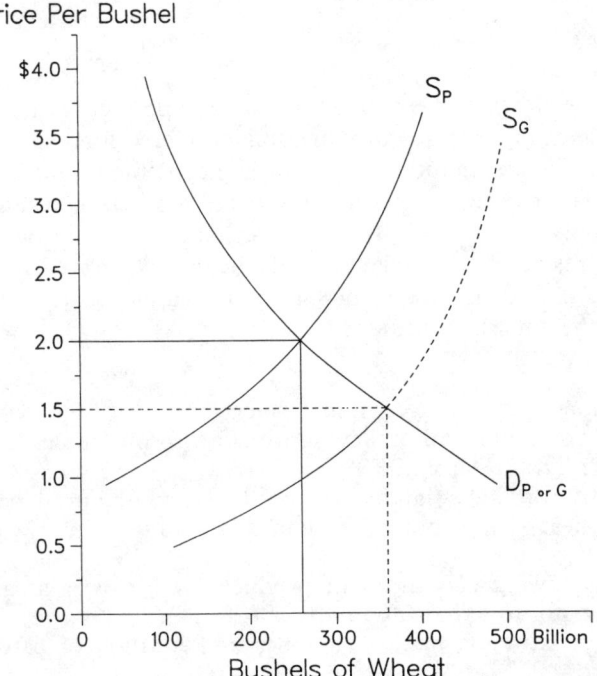

Figure 3.4 Supply and demand for wheat during poor and good years.

price of $2 or the new price of $1.50 are no different during a good year than during a poor year. Willingness to buy (the demand curve) has not been changed by the improvement in weather. Only willingness (and ability) to sell has changed.

Demand Shifters

When the economist draws a demand curve or prepares a tabular demand schedule, he is attempting to describe the changes in product quantity bought due only to changes in price. In other words, he attempts to hold all possible factors constant that might shift the demand curve. Thus, movements along a demand curve represent changes in quantity purchased due only to changes in price, holding the various shifters constant.[8]

What are these shifters of demand? What factors other than product price cause consumers to want to buy more or less of a given product? Demand curve shifters, with specific examples for beef demand, are:

1. Prices of substitutes (lamb, pork, poultry).
2. Prices of complements (hamburger buns, charcoal, red wines).
3. Consumer tastes (fears concerning red meat cholesterol).
4. Population (number of buyers in the market).
5. Per capita disposable income.
6. Consumer expectations concerning prices and income.

Supply Shifters

As with the demand curve, when the economist draws a supply curve he is attempting to describe the changes in quantity of product offered for sale due only to changes in price. The assumption underlying the construction of any supply curve is that all factors that might shift the curve are held constant. Movements along a supply curve, then, trace out changes in quantity of product offered for sale resulting from changes only in product price while holding all shifters constant. Factors capable of shifting the supply curve, with specific examples for the beef market, include:

1. Price of production inputs (forage, grain, labor).
2. Technology and governmental policy [development of Rumensen, federal ban of diethylstilbestrol (DES)].
3. Prices of alternative products (lamb and wool that could be produced on rangeland, wheat that could be produced instead of improved pasture).
4. Weather (forage growing conditions in producer and grower areas, widespread severe storms at calving time).
5. Number of suppliers in the market (number and location of cattle producers and growers).[9]

SUPPLY AND DEMAND — THE BASIC TOOLS

An Application

It may be helpful at this point to apply the concepts of supply and demand as a means of explaining the workings of the United States beef market. Some 15 years ago (January, 1971), the average price of hamburger was $0.49 per pound. By April of 1973, the hamburger price was $0.98, exactly double what it had been only 27 months earlier. What happened during these two years to cause such a drastic increase in price? Was the price rise due to an increase (outward shift) in demand or a decrease (shift backward) in supply, or both?

The price rise is best explained by simultaneous shifts of both curves (Figure 3.5). Supply continued its consistent outward shift during this period but at a smaller than normal rate. Expansion of beef supply was hampered by high feed grain costs, the banning of the growth stimulant hormone DES, and the severe winter and widespread unfavorable weather during the spring calving season of 1973 (Workman, 1975). Concurrent with the less than normal growth of beef supply, there was a greater than usual outward shift of demand. United States population increased only slightly during this period. More important, however, both the world population and per capita income in the United States and throughout western Europe and Japan increased substantially. Meanwhile, adverse weather and expanded United States grain exports

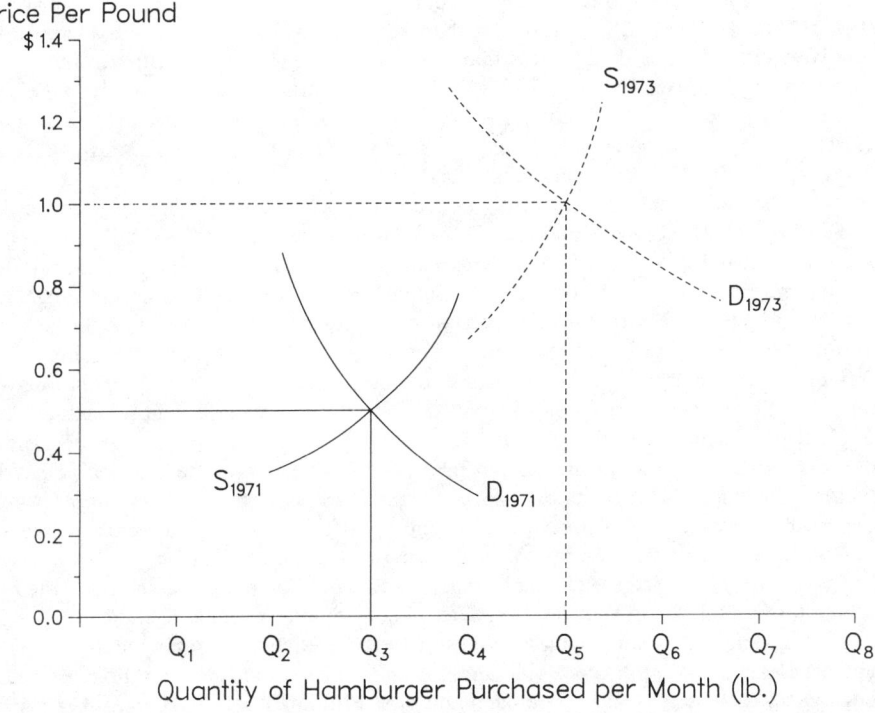

Figure 3.5 Supply and demand curves for hamburger in the United States, 1971 and 1973.

caused feed grain shortages, resulting in rapid increases in the prices of the beef substitutes chicken and pork. This chain of interrelated events created a large increase in the willingness to buy hamburger while restricting expansion of willingness (and ability) to sell beef, resulting in a doubling in the price of hamburger and a less than normal expansion in hamburger sales from Q_3 to Q_5, as shown in Figure 3.5.

APPENDIX I Elasticity of Demand for Beef and Its Effect on Range Improvement Decisions

Price elasticity of demand is a measure of how the amount purchased is affected by changes in the price of a good or service. The elasticity coefficient (E) is defined as the percentage change in quantity purchased divided by the percentage change in product price (Leftwich, 1970). In algebraic terms

$$E = \frac{\Delta Q}{Q} \div \frac{\Delta P}{P} \quad \text{or} \quad \frac{\Delta Q}{\Delta P} \cdot \frac{P}{Q},$$

where Δ denotes "the change in", Q is quantity purchased, and P is product price. If the absolute value of E, $|E|$, is greater than 1, price elasticity of demand is said to be relatively elastic, while if $|E|$ is less than 1, demand is relatively inelastic. In those rare cases when $|E|$ is exactly 1, demand is said to be of unitary elasticity. As the supply of almost any product is increased, price falls, but depending upon the degree of demand elasticity, total *revenue* (price X quantity purchased) will either increase (if $|E| > 1$), decrease (if $|E| < 1$), or remain constant (if $|E| = 1$).

In the latter part of the 1960s two contractory recommendations were issued concerning cattle numbers and the desired level of United States beef production. Numerous articles published in the *Journal of Range Management* recommend that various range improvement practices be implemented in order to increase rangeland cattle production. The recommendations of cattle industry spokesmen during the same period, however, were to reduce beef production by decreasing both the number and weight of cattle marketed. Although contradictory, both types of recommendations are based on the concept of price elasticity of demand. Range management researchers had implicitly assumed that the demand for beef was elastic (i.e., that the beef price decrease resulting from an increase in beef supply would be more than offset by the increase in amount of beef purchased and, therefore, that an expanded supply would bring an increase in total beef sales revenue). Alternatively, cattle industry recommendations must have been based on the implicit assumption that beef demand was inelastic (i.e., that a decrease in beef tonnage sold would be more than compensated by the resulting increase in beef price, also causing beef sales revenue to increase).

In order to determine which of these two contradictory recommendations would actually lead to increased beef sales income and which would not, the price elasticity coefficient for United States beef demand was estimated for the period 1947–1967 (Workman et al., 1972). The resulting coefficient was a relatively inelastic -0.67, indicating that the quantity of beef purchased would decrease by 0.67 percent as a result of a 1 percent increase in beef price. The inverse of the price coefficient, sometimes called the quantity coefficient, Q, is equal to

$$\frac{1}{-0.67} = -1.49$$

and shows that beef price would increase by 1.49 percent in response to a 1 percent decrease in beef quantity. Thus a decrease in beef supply would be more than offset by the corresponding rise in beef price, resulting in more gross income to the beef industry. The recommendation of the cattle industry was the correct one, then, if the goal really was to increase gross[10] beef sales income at the industry level.

Even if maximum gross revenue had been accepted by individual ranchers as their relevant management goal, is it likely that they would have followed the recommendations of industry spokesmen to curtail their cattle sales? The simple "payoff matrix" of Table 3.3 (Richmond, 1957) will be used to help answer this question.

The individual rancher operates in a purely competitive market in which his production is a small, insignificant portion of the total beef supply. Thus he cannot independently influence beef price. Instead he can sell all the cattle he can produce at the going price set in the industry-wide market. The payoff matrix of Table 3.3 is composed of three possible trends in beef supply (willingness to sell) at the industry level and three possible responses by the individual cattle rancher. For illustrative purposes, the three trends in industry beef supply are:

1. Beef supply decreases by 1 percent.
2. Beef supply remains the same.
3. Beef supply increases by 1 percent.

While he cannot independently influence beef supply at the industry level, the individual rancher does control the tonnage of live cattle that he sells. For the sake of simplicity, his three possible courses of action are:

1. Decrease tonnage sold by 1 percent.
2. Continue to sell the same tonnage.
3. Increase tonnage sold by 1 percent.

Table 3.3 Payoff Matrix Showing Nine Possible Changes in Individual Rancher Gross Revenue

Change in live cattle tonnage sold by the individual rancher (%)	Change in industry beef supply (%)		
	−1	0	+1
−1	+0.48	−1.00	−2.48
0	+1.49	0	−1.49
+1	+2.50	+1.00	−0.50

Source: Workman, King, and Hooper (1972).

The outcomes of Table 3.3 are expressed in terms of the combined effects of industry trends and rancher actions on gross income of the individual rancher. For example, if industry beef supply decreased by 1 percent and the rancher chooses to cooperate with the industry trend by decreasing the tonnage of live cattle that he sells by 1 percent, his gross revenue would increase by 0.48 percent over his current earnings.[11]

Based on the assumed rancher goal of maximum gross revenue, the payoff matrix allows selection of the correct rancher course of action corresponding to each of the three trends in industry beef supply. Even without applying the various formal criteria for decision-making under uncertainty (Spencer, 1968), cursory examination of Table 3.3 indicates that no matter which of the three beef industry supply trends the rancher encounters, his correct course of action is to increase his sale of live cattle tonnage by 1 percent. In each of the three industry supply situations, this response either maximizes the increase in gross revenue or minimizes the decrease. The stronger the rancher's expectation that industry beef supply will decrease, the more likely he will increase his own cattle sales. If this expectation is shared by most ranchers, the corresponding rational response will, of course, bring an increase in industry beef supply, rather than a decrease, resulting in a reduction in individual rancher gross revenue. Even if the individual rancher correctly expects the almost certain increase in industry beef supply, an increase in live cattle sales is still the correct course of action for him to minimize his decrease in gross revenue. It seems then, that the cattle industry recommendation to curtail cattle sales was doomed to failure from the beginning.

TEXT NOTES

1 In equation form, our hypothetical demand curve can be expressed as $D = 50 - 25P$ where D is pounds of beef purchased per year, in billions, and P is beef price per pound, in dollars.
2 If the higher beef price prevails for a sufficient time, additional suppliers and producers will enter the beef market by diverting resources away from production and marketing of some other product. The effect of increased price, illustrated in Table 3.2, involves only the increased production and sales offerings of firms already in the beef business and does not include the long term effect of an increase in number of firms.
3 Our hypothetical supply curve can be portrayed mathematically as $S = -6.25 + 31.25P$ where S is pounds of beef, in billions, offered for sale annually and P is beef price per pound, in dollars.
4 The equilibrium price can be determined mathematically by setting the supply equation, $S = -6.25 + 31.25P$, equal to the demand equation, $D = 50 - 25P$, and solving for P (Allen, 1962). Substitution of $P = \$1.00$ into either equation then allows calculation of the equilibrium quantity, 25.
5 It will be recognized, of course, that the beef market, like that for other commodities, is dynamic and subject to constantly changing pressures of weather, consumer tastes, federal legislation, etc. While in the real world beef market, a truly equilibrium price and quantity may never be reached, this does not detract from the usefulness of the concept of equilibrium as the price and quantity that actual price and quantity tend toward. Professor Kenneth Boulding, University of Colorado, has aptly likened actual and equilibrium prices to a dog chasing a

rabbit. While the rabbit (equilibrium price) is where the dog (actual price) would like to be, the dog only moves toward the rabbit, never quite catching it.

6   Consumers are happy in the sense that they are buying exactly as much beef as they want to buy at the price of $1. Of course they would be even happier buying even more beef at an even lower price. Similarly, while producers would be even more satisfied selling even more beef at an even higher price, they are selling exactly as much as they want to sell at the price of $1 per pound.

7   Ruling out changes in consumer expectations, this is true. If wheat consumers expect the wheat crop to be even better next year, they would be wise to postpone their buying until the wheat price is even lower. Such an expectation could shift the demand curve down and to the left.

8   This is no different, of course, than the procedure used by the range manager in attempting to isolate the effects of a single causal factor on, say, reinvasion of unwanted plant species into a reseeded area. If we want to measure the effects of grazing intensity on reinvasion, we would be careful to vary only livestock numbers while holding range site, season of grazing, and duration of grazing constant.

9   This determinant of supply is related to item 3 above.

10   Economic analyses of demand elasticity traditionally treat maximum gross revenue as the relevant management goal. In terms of the average cost curve concept presented in Chapter 5, the implied assumption is that as supply (willingness to sell) is increased or decreased, there is a corresponding change in price but average costs remain constant. It should be recognized, however, that as cattle sales of individual ranches are changed in response to either of these two recommendations, there is a movement either forward or backward along the average cost curve of the individual ranch. Thus in terms of net revenue, which of course is the relevant criterion for the individual rancher, recommendations based on gross revenue could be wrong.

11   Based on the estimated price elasticity coefficient, $E = -0.67$, and the resulting quantity coefficient, $Q = -1.49$, a 1 percent decrease in beef quantity would cause price to increase by 1.49 percent. When the increased beef price is multiplied by the rancher's decreased cattle sales, gross revenue is increased by 0.48 percent: $101.49 \times 0.99 = 100.48$.

LITERATURE CITED

Allen, C. L. 1962. *Elementary Mathematics of Price Theory*. Wadsworth, Belmont, CA. 155 pp.

Leftwich, R. H. 1970. *The Price System and Resource Allocation*. 4th ed. Dryden Press, Hinsdale, IL. P. 39.

Richmond, S. B. 1957. *Statistical Analysis*. 2nd ed. Roland Press, New York. P. 119.

Samuelson, P. A. 1964. *Economics—An Introductory Analysis*. 6th ed. McGraw-Hill, New York. 838 pp.

Spencer, M. H. 1968. *Managerial Economics*. 3rd ed. Irwin, Homewood, IL. 401 pp.

Workman, J. P. 1975. Speculation on future beef production on rangeland. *Rangeman's J.* 2:113–114.

Workman, J. P., S. L. King, and J. F. Hooper. 1972. Price elasticity of demand for beef and range improvement decisions. *J. Range Manage.* 25:337–340.

Chapter 4

The Production Function

PRODUCTION ECONOMICS — THE BASIC PROBLEMS

In production economics theory, *production* is defined as the creation of utility where utility is the usefulness or "want satisfying power" of a good or service that, along with scarcity, gives it value. The various factors used to produce goods and services are usually grouped into the broad categories of land (and other natural resources), labor, capital (funds invested in livestock, machinery, range improvements), and management. The goods and services in question include not only forage, beef, lamb, and wool resulting from the input of rangeland, ranch labor, and various improvements but also the increased after-tax net ranch income and improved soil stability "produced" through consultation with the lawyer, tax accountant, and the range manager.

Production economics theory is directed toward the solution of three basic production problems:

1. *What* should be produced?
2. *How* should it be produced?
3. *How much* should be produced?

These topics are treated in order of increasing complexity, beginning with the

THE PRODUCTION FUNCTION

how much question in Chapter 5, followed by the how question in Chapter 6 and the question of what in Chapter 7. Underlying the solution of each of these problems is the theory of the production function.

THEORY OF THE PRODUCTION FUNCTION

A production function may be defined as the functional relationship between physical (or biological) inputs and physical (or biological) outputs. Although the production function is the basis for all economic analyses concerning how much, how, and what, the function itself is expressed in physical units (acre feet of water, tons of soil) or biological units (pounds of beef, tons of hay, pounds of air dry forage) rather than in dollar terms. The production function is a sort of catalogue of production possibilities showing the various possible combinations of inputs and corresponding outputs. Our examination of the production function will begin with a simplified example where we are concerned with the production of only one product (native meadow hay) using only one fixed input (meadow land) and one variable input (irrigation water).

Table 4.1 Hypothetical Production Function for Meadow Hay

Acre feet irrigation water per acre (A/B)	Tons hay per acre (Y/B)	Tons hay per acre foot (Y/A)	Added tons hay per added acre foot ($\Delta Y/\Delta A$)
Production Stage I			
0.25	0.35	1.40	1.60
0.50	0.75	1.50	1.80
0.75	1.20	1.60	1.60
Production Stage II			
1.00	1.60	1.60	0.80
1.25	1.80	1.44	0.40
1.50	1.90	1.27	0.24
1.75	1.96	1.12	0.20
2.00	2.00	1.00	0.00
2.25	2.00	0.87b	−0.20
Production Stage III			
2.50	1.95	0.78	

Fixed factors of production are defined as those whose use cannot be readily varied when a change in output is desired. In addition to land, other examples include machinery, buildings, and management services under long-term contract. Variable factors are those whose use can be quickly changed in order to bring about a desired change in production. Labor, gasoline, and purchased feeds, along with irrigation water, are examples of variable inputs. Availability of both fixed and variable inputs is usually limited due to actual supply constraints (total amount of meadowland owned or available for lease) or to budget limitations. Despite these various constraints, however, the two types of inputs can be combined in different proportions (by increasing or decreasing the variable input) and, thus, the amount of total product can be varied.

Suppose a western agricultural experiment station sponsored research to determine the relationship between rate of irrigation water application and yield of meadow hay. Application of 10 different irrigation water rates (variable input where A is acre feet of water) on 10 similar 1-acre tracts of meadowland (fixed input where B is acres of meadow) gives the hay yields (output where Y is tons meadow hay) shown in Table 4.1. The data shown comprise a tabular production function showing the biological relationship between intensity of water application and hay yield and illustrating the familiar law of diminishing returns.

LAW OF DIMINISHING RETURNS

The law of diminishing returns states that as a variable input is added to a fixed input, total product first increases at an increasing rate, then increases at a decreasing rate, and finally decreases. In Table 4.1, the variable input (irrigation water) is intensified upon the fixed input (land) resulting in typical changes in total product (hay). Total product is sometimes called *total physical product* of the variable input A, TPP_a, to emphasize that the production function expresses yield in physical (or biological) terms rather than in dollars. The first $0.25A$ applied to $1.00B$ results in the production of $0.35Y$. The next $0.25A$ applied produces $0.4Y$ ($0.75 - 0.35$) and total product is now increasing at an increasing rate. The third $0.25A$ gives an even larger increase in Y of 0.45. Beginning with the fourth $0.25A$, Y begins to increase at a decreasing rate and thereafter each successive increment of water application yields smaller and smaller increases in Y until the tenth increment (the increase from $2.25A$ to $2.5A$) actually brings a decrease in Y.

The law of diminishing returns is even more striking when portrayed graphically (Figure 4.1). The production function (TPP_a curve) becomes steeper and steeper with each successive application of irrigation water up to $0.75A$ after which the TPP curve continues to climb but simultaneously becomes less and less steep until a level plateau is reached between 2.00 and $2.25A$. Beyond $2.25A$, hay production decreases as shown by the loss in TPP_a elevation.

But what of our original question concerning how much water to apply (and how much hay to produce)? By itself, of course, the production function

THE PRODUCTION FUNCTION

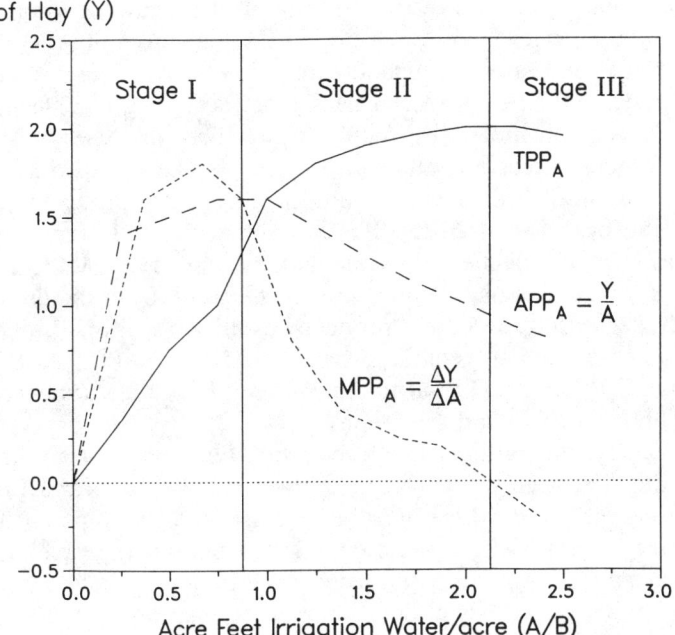

Figure 4.1 Hypothetical production function for meadow hay.

does not provide sufficient information to answer this question. However, the production function does allow a good deal to be said about the rates of irrigation water that should not be applied. For the sake of convenience, since ultimately only a portion of the production function is needed to answer the how much question, the total function may be divided into three distinct stages of production.

STAGES OF PRODUCTION

Identification of the three stages of production requires that two more production measurements be introduced: *average physical product* of the variable input (APP_a) and *marginal physical product* of the variable input (MPP_a). APP_a is defined as total production divided by the number of units of variable input used: TPP/A or Y/A. Thus when $0.25A$ is applied, $APP_a = 0.35/0.25 = 1.40$.

MPP_a is defined as the increased production resulting from application of the last unit of variable input[1]: $\Delta TPP/\Delta A$ or $\Delta Y/\Delta A$ where Δ denotes "the change in." At an application rate of $0.5A$, then, $MPP = (0.75 - 0.35)/(0.50 - 0.25) = 1.60$. Since the calculated value is associated with the change in application of A from 0.25 to 0.50, it is positioned midway between these application rates in both the table and the graph. APP_a and MPP_a values for the meadow hay production function are listed in Table 4.1 and shown graphically in Figure 4.1.

Production Stage I includes the initial portion of the production function up to the variable input level corresponding to the maximum of APP_a ($0.75A$ in Table 4.1). TPP_a increases throughout Stage I, first at an increasing rate (becoming steeper) and then at a decreasing rate (becoming less steep). This phenomenon is best illustrated by MPP_a, which first increases, reaches a maximum, and then decreases to the end of Stage I. The law of diminishing returns is sometimes called the law of diminishing marginal productivity after the characteristic behavior of MPP_a. At the end of Stage I (the boundary between Stages I and II), delineated by the maximum of APP_a, MPP_a is equal to APP_a. Intuitively, as long as the increased Y resulting from the use of one more unit of A, exceeds the average Y produced by all A used ($MPP_a > APP_a$), APP_a must increase. And, of course, whenever $MPP_a < APP_a$, APP_a must decrease. Thus MPP_a equals APP_a at the maximum of the latter (between 0.75 and $1.00A$ in both the table and the graph).

Stage II includes the portion of the production function between the input level where APP_a first begins to decrease (midway between 0.75 and $1.00A$ in the table and graph) and where the TPP_a reaches a maximum (midway between 2.00 and $2.25A$). TPP_a increases at a decreasing rate (becomes less and less steep) throughout Stage II. APP_a decreases throughout Stage II as does MPP_a. However, MPP_a reaches a value of zero at the end of Stage II (the boundary between Stages II and III) and becomes negative in Stage III. Again, intuitively, since MPP_a measures the increased production resulting from the last unit of A applied, MPP_a must equal zero when TPP_a is at a maximum (neither increasing nor decreasing) and MPP_a must take on a negative value in Stage III since TPP is decreasing.

All three parameters decrease throughout Stage III with MPP_a becoming increasingly negative, and both TPP_a and APP_a approaching zero if the application rate of A is sufficiently high.

STAGE II—THE RELEVANT STAGE

The production function provides an accurate measurement of the *efficiency of production*, defined as production per unit of input. TPP_a (tons of hay per acre, Y/B) measures the efficiency of the fixed input (meadowland) and APP_a (tons of hay per acre foot of water, Y/A) measures the efficiency of the variable input (water). For the sake of efficiency, all production must take place somewhere in Stage II and for this reason Stage II is called the relevant stage of production. The rational manager would never produce in Stage III since the efficiency of the fixed input (as measured by TPP_a) has been pushed past its maximum and production of Y could be increased by applying less A. Even if the variable input, irrigation water, could be obtained free (allowing us to ignore the efficiency of water and to instead concentrate solely on the efficiency of the fixed input, land), we would never apply sufficient water to depress hay yields.[2]

THE PRODUCTION FUNCTION

For similar, but somewhat more complicated reasons, production should never take place in Stage I. The efficiency of the variable input (as measured by APP) has not yet reached a maximum in Stage I and Y could be increased by applying more water. Suppose we were producing in Stage I of our meadow hay example. It is fine to know that maximum efficiency of the variable input has not been reached and that hay production could be increased by increasing our water application rate. We would like to reach Stage II and thus increase our hay production (in fact we would like to have sufficient free water to push our application rates to the boundary between Stages II and III and thus reach maximum hay yields from our fixed quantity of meadowland). But what if we are already applying all of our limited (and costly) variable input, water, to our fixed input, land? How can we reach Stage II and the resulting increase in hay yield?

The law of diminishing returns is sometimes called the law of variable proportions since it is the change in the ratio or proportion of variable input to fixed input (A/B) that actually causes the increases (and decreases) in production efficiency as the variable input application rate is changed. We prefer more hay from our land and water to less hay and we need to find a means of increasing the water/land ratio. Since our variable input, water, is limited and since we are already applying the total available, the adjustment in the A/B ratio must be made by changing the quantity of land. Although land is our fixed input in the sense that we cannot buy or lease additional meadowland on short notice, the use of our total available land is flexible, and our water/land ratio can be increased by cutting back on the amount of land watered. Thus we can reach Stage II (and, more importantly, we can produce more hay) by allowing part of our meadowland to go unwatered. Great theory! But will it really help us increase our hay production? Our fixed input, land, costs a bundle each year in taxes and interest and we are naturally somewhat reluctant to allow a portion of it to lie idle in the hope that our hay production will be increased on the land that is watered.

Let us put the variable proportions theory to a test. The test will be confined to 1 acre and 0.25 acre feet of water from Table 4.1. The relationships developed for this small parcel and small amount of water will hold true for any amount of similar meadowland. If the 0.25 acre feet of water is spread evenly over the entire acre of land (an A/B ratio of $0.25:1$), 0.35 tons of hay will be produced. Now suppose that our limited water is applied to only one-half of our land and the remaining one-half acre is allowed to lie idle. The $0.25A$ applied to $0.5B$ gives a water/land ratio of $0.5:1$. This same proportion of A to B gave a per acre hay yield of 0.75 tons. Thus our half acre produces 0.375 tons, or 0.025 tons more hay than if the limited water were applied to the entire acre. This is a definite improvement but perhaps we can do even better. As shown in Table 4.2, application of the limited irrigation water on only one-third of the available land (an A/B ratio of 0.75) further increases total hay production and, as shown by the fact that water efficiency (Y/A) has

Table 4.2 Hypothetical Production Function for Meadow Hay When a Limited Quantity of Irrigation Water is Applied to Various Portions of 1 Acre.

Portion of the acre lying idle	Portion of the acre irrigated	Acre feet irrigation water per acre (A/B)	Tons hay per acre (Y/B)	Tons hay per acre foot (Y/A)	Total tons hay
Production Stage I					
0	1.00	0.25	0.35	1.40	0.350
0.50	0.50	0.50	0.75	1.50	0.375
0.67	0.33	0.75	1.20	1.60	0.400
Production Stage II					
0.75	0.25	1.00	1.60	1.60	0.400
0.80	0.20	1.25	1.80	1.44	0.360

reached a maximum, has simultaneously allowed us to reach the boundary between Stages I and II. The 0.4 tons of hay is the maximum that can be produced by any combination of 0.25 acre feet of water and 1 acre of land (or portion thereof). Thus, the variable proportion theory has passed our test. It seems we really can reach Stage II by increasing our water/land ratio through reduction of the amount of land used. Even more important, by varying the proportions of limited variable and fixed inputs used, we have reached our more pragmatic goal of increasing total hay production.

Logically then, even if the fixed land input were free (and unlimited supplies were available), allowing us the luxury of concerning ourselves only with the efficiency of the limited and costly variable water input, we should never use so much land in combination with our limited water that we operate in Stage I. Instead we should apply the limited water to just enough unlimited land to maximize water efficiency (APP_a). In other words, even if land were free, we should still attempt to operate at the boundary between Stages I and II.[3]

In summary, Stage II is the relevant production stage and no matter what product, variable inputs, and fixed inputs we are concerned with, we should always strive to operate somewhere in this stage. Where in Stage II we should operate (the "how much" question) is the subject of Chapter 5 and can be determined only after the production function is analyzed in combination with the prices of the product and all inputs. However, even at this point it can be stated with absolute certainty that production should never take place to the right of the boundary of Stages II and III nor to the left of the boundary of Stages I and II. Even if the variable input were free the rational producer would apply only enough to reach the boundary between Stages II and III. Alternatively, even if the fixed input were free and unlimited it would be highly irrational to combine any more of it with the limited variable input than

THE PRODUCTION FUNCTION

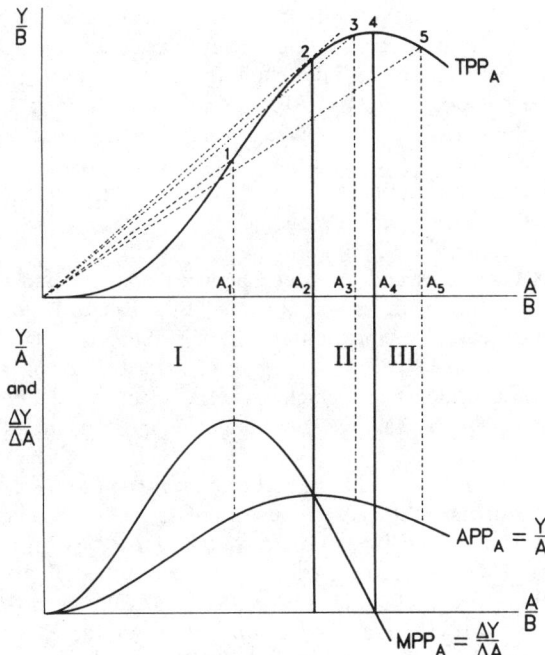

Figure 4.2 Derivation of APP_a and MPP_a from TPP_a.

exactly enough to push production to the left so as to just reach the boundary between Stages I and II.

APPENDIX I The General Production Function Model

In the above irrigation water example, TPP_a, APP_a, and MPP_a curves were plotted by joining a few data points with straight-line segments. In general form, the TPP_a function appears as a continuous, smooth sigmoid curve (Figure 4.2). Up to a point 1, the TPP_a curve increases at an increasing rate (becoming more and more steep) causing it to be convex. Between points 1 and 4 the curve continues to increase but at a decreasing rate (becoming less and less steep) causing it to be concave. Beyond point 4 TPP_a actually decreases (turns down) and continues to be concave, completing its sigmoid or S shape. The characteristic S shape of the TPP_a curve, then, like the shape of all product and cost curves discussed in this book, is due to the now familiar law of diminishing returns.

Both the APP_a and MPP_a curves can be derived directly from TPP_a. Since APP_a represents TPP/A, APP_a for any particular quantity of variable input A can be calculated as the slope (vertical distance Y divided by horizontal distance A or "the rise over the run") of a straight-line segment or ray from the origin to the point on the TPP_a curve corresponding to the particular level of A in question.[4] Even without using specific numbers (the axes of the graphs of Figure 4.2 lack numerical scales), the general form of APP_a can be derived. Since TPP_a begins at the origin of the graph, the APP_a curve must also begin at the origin. Between the origin and point 2, the slope of rays from the origin to various locations along the TPP_a curve (point 1, for example) becomes ever steeper. Thus between the origin and A_2, APP_a is increasing. Beyond A_2 the slope of rays drawn to points along TPP_a becomes successively less steep. The ray to point 2, then, is the steepest straight line that can be drawn out of the origin and still touch TPP_a, so APP_a must reach a maximum at A_2. Beyond A_2, APP_a becomes smaller and smaller, approaching but never actually reaching zero.[5]

MPP_a traces out $\Delta Y/\Delta A$ or the slope of TPP_a at a specific point.[6] Since TPP_a becomes ever steeper up to point 1 and less and less steep between points 1 and 4, MPP_a must reach a maximum at A_1, the inflection point, where TPP_a changes from a curve that is convex to one that is concave. The maximum or peak of TPP_a is reached at point 4 and at this point $MPP_a = 0$ since TPP_a has a slope of zero. MPP_a can be more strictly defined as the slope of a straight-line segment drawn tangent to the TPP_a curve. The ray between the origin and point 2 is one such tangency and thus at A_2, MPP_a is equal to APP_a and cuts it from above.

As already mentioned in the discussion of the discrete number irrigation water example above, as long as MPP_a exceeds APP_a, APP_a must increase and vice versa. Thus the boundary between Stages I and II corresponds to A_2 where $APP_a = MPP_a$. The boundary between Stages II and III occurs at A_4 where $MPP_a = 0$.

In summary, APP_a and MPP_a curves can be easily derived geometrically from the general TPP_a curve. The derived curves provide a fast and accurate means of identifying Stage II, the relevant stage that forms the basis of the analysis of Chapter 5.

FURTHER READING

Allen, C. L. 1962. *Elementary Mathematics of Price Theory*. 1st ed. Wadsworth, Belmont, CA. 155 pp.
Ferguson, C. E. 1966. *Microeconomic Theory*. 1st ed. Irwin, Homewood, IL. 439 pp.
Gisser, M. 1966. *Introduction to Price Theory*. 1st ed. International Textbook Co., Scranton, PA. 325 pp.
Heady, E. O. 1952. *Economics of Agricultural Production and Resource Use*. Prentice-Hall, Englewood Cliffs, NJ. 850 pp.
Leftwich, R. H. 1970. *The Price System and Resource Allocation*. 4th ed. Dryden Press, Hinsdale, IL. 402 pp.
Samuelson, P. A. 1964. *Economics—An Introductory Analysis*. 6th ed. McGraw-Hill, New York. 838 pp.

TEXT NOTES

1 In mathematical terms, $MPP_a = dY/dA$ where $Y = f(A|B)$, A is variable input, and B is fixed input.
2 In actual irrigation practice, the rule of stopping short of Stage III is usually followed. Even those landowners with abundant water are careful not to drown their native meadows if topography and canal systems allow sufficient water control.
3 Unlike the natural tendency to avoid Stage III described above, the rule of pushing the production process out of Stage I and into Stage II by allowing some land to go

unwatered is often violated in actual irrigation practice. The counterintuitive logic underlying the rule may be the reason.
4 Mathematically, $APP_a = Y/A$ where $Y = TPP_a = f(A|B)$, A is variable input, and B is fixed input.
5 In mathematical terms, APP_a approaches the horizontal axis asymptotically.
6 Mathematically, $MPP_a = dY/dA$ where $Y = f(A|B)$, A is variable input, and B is fixed input.

Chapter 5

Optimum Intensity of Production

This chapter deals with the "how much to produce" question posed in Chapter 4. It should be clear from the Chapter 4 discussion that whenever a variable input is being used in combination with a fixed input to produce some product, the optimum input level will lie somewhere in production Stage II. Exactly where in Stage II production should take place (and whether or not we should produce at all) depends upon prices: the price of the product, the price (and resulting cost) of the variable input, and the price (and resulting cost) of the fixed input. Determination of the optimum intensity or scale of production is based upon the concept of "marginality."

THE MARGINALITY PRINCIPLE

A hypothetical production function for weight gains of steers (variable input) grazing rangeland (fixed input) at various stocking rates for a two month season is shown in Table 5.1. The effects of the law of diminishing returns are evident. An increase in stocking rate from 5 to 10 steers per section (640 acres) brings an increase in steer gain of 500 pounds and more than doubles the initial 5 steer production of 400 pounds. This increment represents the "increasing at an increasing rate" or convex portion of the TPP function and

Table 5.1 Hypothetical Production, Costs, and Returns From Steers Grazing Rangeland

Variable input (steers)	Fixed input (acres)	Total product (TPP) (pounds steer gain)	Total revenue (TR) ($)	Marginal revenue (MR) ($)	Total fixed cost (TFC) ($)	Average fixed cost (AFC) ($)	Total variable cost (TVC) ($)	Average variable cost (AVC) ($)	Total cost (TC) ($)	Average total cost (ATC) ($)	Marginal cost (MC) ($)	Net revenue (NR) ($)
5	640	400	200		100	0.25	100	0.25	200	0.50	0.20	0
				0.50								
10	640	900	450		100	0.11	200	0.22	300	0.33	0.40	150
				0.50								
15	640	1150	575		100	0.09	300	0.26	400	0.35	0.48	175
				0.50								
20	640	1360	680		100	0.07	400	0.29	500	0.36	1.00	180
				0.50								
25	640	1460	730		100	0.07	500	0.34	600	0.41	2.50	130
				0.50								
30	640	1500	750		100	0.07	600	0.40	700	0.47	-5.00	50
35	640	1480	740		100	0.07	700	0.47	800	0.54		-60

might be explained by the steers walking the fence searching for other cattle when only 5 are present on a section but settling down and grazing when steer numbers are doubled. Increased stocking rates in 5 steer increments from 10 to 30 steers bring increased steer gains per section, but the increase occurs at an ever decreasing rate. The increase from 10 to 15 steers allows an additional gain of 250 pounds per section but the next 5 steer increase from 15 to 20 steers results in only 210 pounds of gain. Each successive 5 steer increase yields smaller additions to total steer gain, due to ever increasing competition for forage. Total forage consumption and resulting total steer gain both increase but at the expense of reduced gains per steer causing the rate of increase to decline. In the extreme, an increase from 30 to 35 steers per section causes forage competition to become so severe that total steer gain per section declines.

Marginal Revenue

Total steer gain is transformed to *total revenue* (TR) by simply multiplying pounds of gain by the price per pound ($0.50 in this example).[1] *Marginal revenue* (MR) is defined as the additional revenue produced by the last unit of output or MR = $\Delta TR/\Delta Y$ where Y is output (steer gain).[2] Thus MR between stocking rates of 5 and 10 steers is ($450 − $200)/(900 − 400) = $0.50.[3] In the purely competitive market for cattle, the rancher can buy or sell all the steers he wants at the going market price of $0.50 per pound. His purchases or sales are too small a portion of total sales to alter market price and for this reason MR for steer gains in Table 5.1 remains constant at $0.50 per pound.

Fixed versus Variable Costs

Production costs are conventionally separated into fixed costs and variable costs. *Fixed costs* are those attributable to the fixed factors of production and include both cash costs such as land rent (paid on an acreage basis), property tax, insurance, and opportunity costs such as operator and family labor and interest on investment. Fixed costs do not change with a change in output. They do not increase with increased production nor can we avoid paying these costs in the "short run"[4] by ceasing our operation. In the simple example of Table 5.1, fixed costs are comprised of only an annual minimum land charge of $100. This is the minimum service fee charged by the land owner to initiate a grazing lease and this amount, designated total fixed cost (TFC) in Table 5.1, must be paid no matter how many steers are grazed per section. This fee must be paid at the outset of the lease agreement and is not refundable even if the cattle owner decides later to not run any steers. Average fixed cost (AFC) is TFC divided by output, TFC/Y = TFC/TPP. AFC decreases as stocking rate is increased since TFC is spread over ever increasing steer gains.

Variable costs are those associated with the variable factors of production such as purchased feed, repairs, gasoline, and hired labor. Variable costs do increase with increases in output and can be thought of as costs that we can

OPTIMUM INTENSITY OF PRODUCTION

avoid paying by ceasing our operation. In Table 5.1, variable costs are composed of grazing fees (paid in addition to the fixed minimum service fee of $100 per year) charged at the rate of $10 per steer per month. Total variable cost (TVC) at a stocking rate of 5 steers per section is 5 steers × $10 × 2 months = $100. Average variable cost (AVC) is TVC divided by output, TVC/Y = TVC/TPP. The reader will recall that the shape of the TPP function is due to the law of diminishing returns. This same law causes AVC to at first decrease with heavier stocking rates (up to 10 steers per section) and to then increase (from 10 to 35 steers per section).

Total cost (TC) is simply the sum of TVC and TFC. TC represents all costs associated with any particular level of output. Average total cost (ATC) is TC divided by output, TC/Y. Again, due to the law of diminishing returns, ATC decreases at first with stocking rate increases (up to 10 steers per section) and then increases (from 10 to 35 steers per section).

Marginal Cost

Marginal cost (MC) is defined as the additional cost attributable to the last unit of output or MC = $\Delta TVC/\Delta Y$ or, since the difference between TVC and TC is the constant TFC, MC = $\Delta TC/\Delta Y$.[5] At a stocking rate of 10 steers per section MC = ($100 − $100)/(900 − 400) = $0.10. As with the MR values mentioned above, $0.20 is the MC associated with the increase in output from 400 to 900 and the $0.20 should be placed halfway between 400 and 900 when MC values are graphed or arranged in a table. MC continues to increase throughout the range of output shown in Table 5.1 and again, due to the law of diminishing returns, the increase occurs at an increasing rate.[6]

Equating at the Margin for Optimum Intensity

If we are attempting to meet the usual management goal of profit maximization and if unlimited capital is available for expansion of our operation, the optimum level of production (optimum stocking rate) is the one where MC = MR.[7] In Table 5.1, then, the optimum stocking rate is 20 steers per section since more net revenue is earned at this stocking rate than at any other. This 20 steer stocking rate where costs and returns are "equated at the margin" can be thought of as the level of production where the last unit of output (steer gain) adds the same amount of costs as to revenue.[8] The intuitive logic underlying the equating of MR and MC is simple. Between 10 and 15 steers per section, additional steer gain from intensification of stocking rate adds more to revenue than to costs (MR = $0.50 > MC = $0.40) and it pays the manager to increase the number of steers run. Between 20 and 25 steers per section, the additional steer gain adds more to costs than to revenue (MC = $1.00 > MR = $0.50) and if current stocking rate were 25 steers per section. net revenue could be increased by running less steers. Only at the optimum stocking rate of 20 steers per section is there no incentive to either increase or decrease steer numbers. Since MR = $0.50 is approximately equal to MC =

$0.48 we know with certainty (as verified by the net revenue data of Table 5.1) that 20 steers per section is the optimum stocking rate to maximize net revenue.

Two Common Fallacies

Due perhaps to poor communication between economists and range managers, two erroneous beliefs concerning optimum grazing intensity are widely held by range managers (and by biologists in general, including animal scientists and agronomists). These fallacies often find their way into range management literature.

The "Blame the Profit Motive" Fallacy This first fallacy involves the idea that it is the profit maximization goal of the private enterprise system that is responsible for excessively heavy stocking rates and resulting deterioration of privately owned rangelands.[9] The fear is sometimes expressed that if the private market system is allowed to function unhampered, the search for profits will lead to irreversible degradation of private rangelands. While ignorance (and optimism) concerning rangeland carrying capacity has damaged much private rangeland, this damage has occurred in spite of the profit motive, not because of it.

The hypothetical steer gains of Table 5.1 represent a sustained yield production function such as might be estimated by long-term grazing intensity research (Workman and Lacey 1982). Thus, we can be confident that 30 steers can be grazed year after year on 640 acres for 2 months and an average gain (with wide fluctuations due to weather) of 1500 pounds will result. A stocking rate of 35 steers per section, however, will lead to an average gain of only 1480 pounds. The sustained yield decrease is due to forage competition among animals and to long-term damage to vegetation and soil.

The hypothetical sustained yield relationship between stocking rate and total steer gain is presented graphically in Figure 5.1. Given this production function, the stocking rate recommended by the biologist has commonly been to maximize sustained yield or, as in Figure 5.1, to run 30 steers per section. The rationale is that to run fewer than 30 steers would waste forage that could be converted to a meat product useful to man. Running more than 30 steers, however, reduces total steer gain and causes long-term range damage.

But, as noted above, the stocking rate recommended by the economist (based on the maximum profit motive of the private land owner) is only 20 steers per section. Thus, the profit motive leads to a lower stocking rate than maximum sustained yield and one that poses less risk of range degradation than the biologist's recommendation. While the economist (and in most cases private landowners as well) share the biologist's wish to alleviate hunger and avoid the waste of forage, it simply costs more for the additional steer gain produced by going beyond 20 head than society is willing to pay. The

OPTIMUM INTENSITY OF PRODUCTION

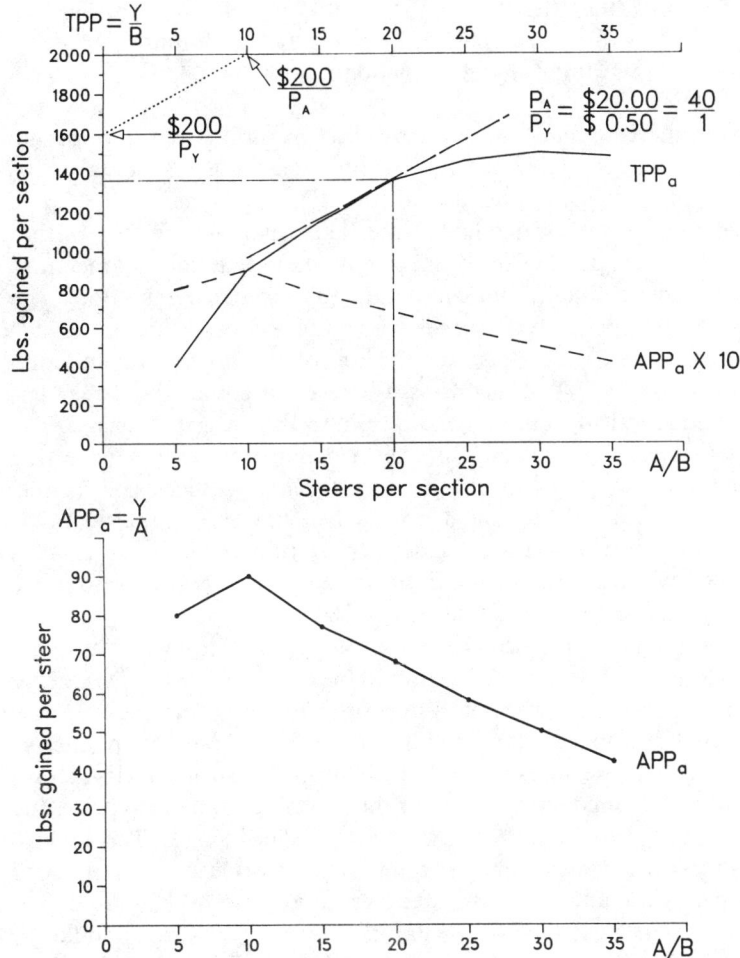

Figure 5.1 Hypothetical production function, inverse price line, and optimum stocking rate for steers grazing 640 acres of rangeland.

economist's recommendation recognizes that scarce operating capital must be spent where it will do the most good (as measured by the value of the product produced), which, at the prices in the example, rules out stocking rates beyond 20 steers. It should be noted that in general the only market condition under which the economic optimum (maximum profit) would correspond to maximum sustained yield would be if marginal cost were zero (in other words, if additional steers could be run at zero extra cost).

The top portion of Figure 5.1 demonstrates the optimization procedure of Table 5.1 in graphical form. The graphical approach shown answers the question of optimum intensity in terms of optimum intensity of input (number of steers per section) rather than from the standpoint of optimum level of output (steer gain). As will be demonstrated, the input and output approaches both lead to the same stocking rate recommendation.

The "Mysterious Optimum" fallacy An understanding of the optimum input approach of Figure 5.1 is required to appreciate the reasoning errors involved in the "mysterious optimum" fallacy. The line labeled P_A/P_Y in Figure 5.1 is called an inverse price line. Note that it is a straight line with a slope equal to the inverse price ratio, P_A/P_Y or 40.00,[10] that comes tangent to the TPP curve at the optimum stocking rate of 20 steers per section. The correct position and slope of the inverse price line are easily determined.[11] First, another horizontal axis is drawn at the top of the production function graph as shown in Figure 5.1. This forms a second origin in the upper left corner of the graph identical to the original origin in the lower left corner. The new horizontal axis is scaled exactly like the original, measuring increasing A/B (steers/section) from left to right. Second, the new vertical axis formed by the upper left graph origin is scaled exactly like the original vertical axis except that it is upside down and increasing Y/B (pounds gain/section) is measured along the vertical axis from top to bottom. Next, two questions are asked. First, if we had an operating budget of $200[12] and spent it all on steer gain, how many pounds of gain could we buy? The answer, budget = $200/$0.50 per pound = 400 pounds, is plotted on the vertical axis, measuring from top to bottom. Second, if we spent the entire $200 on grazing leases, how many steers could we run for the 2-month season? The answer, budget = $200/$20 per steer = 10 steers, is plotted on the upper horizontal axis. Third, the two points are connected with a straight line forming an inverse price line with a slope of $(1/P_Y)/(1/P_A) = P_A/P_Y = \$20/\$0.50 = 40/1$. The inverse price line or isocost (equal-cost) line represents all possible combinations of purchases of steer gain and 2-month steer grazing leases that would just exhaust a $200 budget. Next the isocost line is moved downward, parallel to the initial $200 isocost line, until it comes tangent to the TPP curve. This tangency marks the optimum stocking rate at 20 steers per section resulting in the production of 1360 pounds of steer gain per section, identical to the optimum grazing intensity determined in Table 5.1.

The tabular and graphical approaches of determining optimum intensity are mathematically identical. By definition, the slope of the TPP curve in Figure 5.1, at any particular point on the curve, is equal to MPP at that point; as shown above, the slope of the isocost line is equal to P_A/P_Y. Thus, where the isocost line comes tangent to the TPP curve the following equality holds:

$$\text{MPP}_A = \frac{P_A}{P_Y} \quad \text{or, rearranging terms}$$

$$P_Y = \frac{P_A}{\text{MPP}_A}.$$

OPTIMUM INTENSITY OF PRODUCTION

By definition, $P_Y = \Delta TR/\Delta Y = MR$, $P_A = \Delta TC/\Delta A$, $MPP_A = \Delta Y/\Delta A$, and $\Delta TC/\Delta Y = MC$. Thus the above equality can be written

$$MR = P_Y = \frac{P_A}{MPP_A} = MC = \frac{\Delta TC/\Delta A}{\Delta Y/\Delta A} = \frac{\Delta TC}{\Delta Y} = MC \text{ or}$$

$$MR = MC$$

and the mathematical relationships specifying the optimum input level in Figure 5.1 are identical to those of Table 5.1.

We are now equipped to examine the "mysterious optimum" fallacy concerning the optimum stocking rate. Based on exhaustive surveys of range management literature, both of the most recent and most widely used range management textbooks [*Range Management* by Stoddart, Smith, and Box (1975) and Heady's (1975) *Rangeland Management*] correctly state that the optimum stocking rate lies somewhere between the grazing intensity giving maximum gains per animal (10 steers per section in the example of Figure 5.1) and the intensity yielding maximum gains per acre (30 steers per section in Figure 5.1). Without actually using economic terminology, both textbooks correctly conclude that the optimum stocking rate lies somewhere between maximum APP and maximum TPP or somewhere in production Stage II. Both textbooks also conclude that many grazing intensity studies that have been conducted to determine the optimum stocking rate have proven inconclusive for a variety of reasons (insufficient range of stocking rates to measure crucial vegetation and livestock responses, insufficient seasons studied to detect ultimate vegetation and soil changes, etc.).

Precise determination of the optimum stocking rate is difficult in any range grazing situation. Unfortunately, however, both of the above range management textbooks (and numerous other grazing intensity publications) also leave the erroneous impression that the true optimum stocking rate is also difficult to even define. The reader is left with the idea that about all that can be said about this "mysterious optimum" stocking rate is that it lies somewhere between maximum per animal performance and maximum per acre performance, or, to use the common but less precise term, moderate grazing is the economic optimum rate. Despite a clear explanation of the application of economic concepts to the determination of optimum grazing intensity many years ago by Hopkin (1957) and empirical examples by Hildreth and Riewe (1963), these concepts have not been widely adopted in either range science research or in grazing practice. Notable exceptions are publications by Gray (1968), Bement (1969), Schoop and McIlvain (1971), Pearson (1973), and Whitson and Ragsdale (1976) which do correctly specify the optimum stocking rate in terms of maximum net returns per acre.[13]

A more sophisticated version of the "mysterious optimum" fallacy, but a fallacy just the same, involves the notion that the optimum stocking rate is the one where the production per animal (APP) curve intersects the production per acre (TPP) curve (Stoddart, 1960; Roberts, 1979). This idea may have come about because of the range manager's legitimate concern that the maximum sustained yield stocking rate (maximum TPP) often recommended (or at least alluded to in range management literature) may pose excessive risk of resource degradation in the event of a drought or overestimation of carrying capacity, or this notion may have resulted from the work of Mott (1960) who *arbitrarily* specified an optimum stocking rate and then scaled his graph axes in terms of ratios of actual production to optimum production. Whatever the cause, the definition of optimum stocking rate as the one where APP = TPP is incorrect. This fact is easily demonstrated in Figure 5.1 above. The APP curve, derived from TPP in Table 5.1, is plotted in both the lower and upper graphs at a much larger scale than that used for TPP in the upper graph. If the APP were plotted in the upper graph at the same scale used to trace out TPP, it would lie far below TPP and would not intersect it anywhere. In order to obtain an intersection, APP has been plotted in the upper graph at a scale 10 times the correct scale of the lower graph. APP does cross TPP in the upper graph at a stocking rate of 10 steers per section but of course there is absolutely no significance in where the two curves intersect. The correct scale was arbitrarily increased by a factor of 10 to force APP to intersect TPP at this particular stocking rate and APP can be made to cross TPP at any desired stocking rate by simply manipulating the vertical scale.

In summary, one of the most important problems in managing private rangelands (and one that the range management profession is looked to for guidance) is the determination of the economic optimum stocking rate. As long as there is any cost associated with the variable input (numbers of livestock), the economic optimum (maximum profit) stocking rate is always less than maximum sustained yield.[14] Thus, the economic optimum poses less risk of resource degradation than the biological optimum of maximum sustained yield.[15] The economic optimum also has nothing to do with equating per animal and per acre productivity. Instead, since the economic optimum stocking rate depends upon the cost of the variable input and the price of the product, it is the proven economic concept of marginality that is crucial to its determination. The stocking rate problem is sufficiently important to warrant the use of this superior quantitative tool.

Factors Affecting Optimum Intensity and Profit

The optimum stocking rate of 20 steers per section and resulting maximum net revenue of $180 in Table 5.1 were calculated for specific input and output prices. The factors affecting optimum intensity and profit (net revenue) include (1) price of the product, (2) price of the variable input, (3) price of the fixed input, and a related factor (4) the conversion of a variable input to a fixed input or vice versa. Those factors are described in sequence below.

Product Price In Table 5.1 the product price is $0.50 per pound of steer gain. How would profit and optimum stocking rate be affected by a decrease in

OPTIMUM INTENSITY OF PRODUCTION

Table 5.2 Lowered Optimum Stocking Rate and Decreased Net Revenue Due to a Decline in Product Price

Variable input (steers)	Fixed input (acres)	Total product (TPP) (pounds steer gain)	Total revenue (TR) ($)	Marginal revenue (MR) ($)	Total fixed cost (TFC) ($)	Total variable cost (TVC) ($)	Total cost (TC) ($)	Marginal cost (MC) ($)	Net revenue (NR) ($)
5	640	400	168	0.42	100	100	200	0.20	−32
10	640	900	378	0.42	100	200	300	0.40	78
15	640	1150	483	0.42	100	300	400	0.48	83
20	640	1360	571	0.42	100	400	500	1.00	71
25	640	1460	613	0.42	100	500	600	2.50	13
30	640	1500	630	0.42	100	600	700	−5.00	−70
35	640	1480	622		100	700	800		−178

livestock price? Suppose steer prices fell to $0.42 per pound. Would profit go up or down? Based solely on common sense, most persons would correctly answer that a reduced steer price would bring a reduction in profit. But what about the effect on optimum stocking rate? The resulting change is not so obvious and common sense is less helpful. Many people are tempted to approach the question intuitively (and incorrectly) as follows: If steer price falls, the ranch operator should attempt to maintain his income by running more steers. This reasoning has led to the false belief among some range managers that lower cattle prices bring heavier stocking and vice versa. As shown in Table 5.2, the correct stocking rate adjustment to the lowered steer price is exactly the opposite. At the $0.42 steer price, the optimum stocking rate drops to 15 steers per section.[16] Maximum profit also decreases from $180 to $83. Still, the 15 steer optimum yields more net revenue than any other possible stocking rate as shown by the last column of Table 5.2. Counter to intuitive logic, the ranch operator clearly cannot afford not to cut back on stocking rate. This example demonstrates that whenever product price (MR) drops, optimum production intensity and net revenue both decrease and vice versa.

Price of Variable Input In the original situation of Table 5.1, the cost of additional units of variable input (steers) was calculated in terms of a variable grazing fee of $10 per steer per month or a total variable cost of $20 per steer for the 2-month grazing season. Now suppose that the lease fee is increased to

Table 5.3 Lowered Optimum Stocking Rate and Decreased Net Revenue Due to an Increase in Variable Grazing Fee

Variable input (steers)	Fixed input (acres)	Total product (TPP) (pounds steer gain)	Total revenue (TR) ($)	Marginal revenue (MR) ($)	Total fixed cost (TFC) ($)	Total variable cost (TVC) ($)	Total cost (TC) ($)	Marginal cost (MC) ($)	Net revenue (NR) ($)
5	640	400	200		100	120	220		−20
				0.50				0.24	
10	640	900	450		100	240	340		110
				0.50				0.48	
15	640	1150	575		100	360	460		115
				0.50				0.57	
20	640	1360	680		100	480	580		100
				0.50				1.20	
25	640	1460	730		100	600	700		30
				0.50				3.00	
30	640	1500	750		100	720	820		−70
				0.50				−6.00	
35	640	1480	740		100	840	940		−200

$12 per steer per month while all other prices remain as they are in Table 5.1. Would profit increase or decrease? Most persons would intuitively answer correctly that an increased grazing fee would decrease profit, as verified in Table 5.3. But what impact will this variable cost change have on optimum stocking rate? Should steer numbers be increased or decreased in response to the higher variable grazing fee? Many people find that the correct answer runs counter to intuitive reasoning and they may be tempted to argue (incorrectly) that stocking rate should be increased to compensate for increased costs in hopes of maintaining as much net revenue as possible. Again, as shown in Table 5.3, the correct stocking rate adjustment is just the opposite. The increase in variable cost brings a corresponding increase in marginal cost and equality of MC and MR comes at a lower stocking rate of 15 steers per section. Profit is also decreased, of course, from $180 to only $115 per section. But more profit is earned with 15 steers than with any other possible stocking rate.[17] This example demonstrates the general conclusion that an increase in the price of a variable input will always bring a decrease in both optimum production intensity and profit while variable cost decreases will have the opposite effect.

Price of Fixed Input The cost of the fixed input in Table 5.1 was $100 per section of rangeland for the entire grazing season. This amount represented a minimum service fee charged by the landowner for initiating a grazing lease. Suppose that the leasor waived the fixed lease fee of $100 per section. If all

OPTIMUM INTENSITY OF PRODUCTION

Table 5.4 Unchanged Optimum Stocking Rate and Increased Net Revenue Due to a Decrease in the Fixed Costs of the Grazing Lease

Variable input (steers)	Fixed input (acres)	Total product (TPP) (pounds steer gain)	Total revenue (TR) ($)	Marginal revenue (MR) ($)	Total fixed cost (TFC) ($)	Total variable cost (TVC) ($)	Total cost (TC) ($)	Marginal cost (MC) ($)	Net revenue (NR) ($)
5	640	400	200	0.50	0	100	100	0.20	100
10	640	900	450	0.50	0	200	200	0.40	250
15	640	1150	575	0.50	0	300	300	0.48	275
20	640	1360	680	0.50	0	400	400	1.00	280
25	640	1460	730	0.50	0	500	500	2.50	230
30	640	1500	750	0.50	0	600	600	-5.00	150
35	640	1480	740		0	700	700		40

other prices remain at the original levels of Table 5.1, how will the $100 reduction in fixed cost affect profit? Based on intuition alone, most persons would correctly answer that the reduced fixed cost would increase profit. But what effect would the decreased fixed cost have on optimum stocking rate? As with the other price changes mentioned above, many people are tempted to extend their common sense logic by arguing (incorrectly) that since costs have been reduced, the cattle owner can now afford to cut back on stocking rate. However, as shown in Table 5.4, the reduction in fixed costs has no effect on MC and the optimum stocking rate remains unchanged at 20 steers per section. Since total cost is reduced by $100 at all output levels, profit is increased by $100 at the optimum rate as well as at all other intensities of range use.

This example demonstrates that a decrease (or increase) in price of the fixed input has no effect on optimum production intensity but does increase (or decrease) profit. Thus fixed costs have no bearing on optimum level of output except in the extreme "whether to produce at all" decision discussed under the heading of "The General Marginality Model". In this extreme case, if the fixed cost increase is sufficient to more than offset all profit, the appropriate action is to cease production.

Conversion of a Variable Cost to a Fixed Cost Opportunities are often available on a range livestock operation to convert what is currently a variable cost to a fixed cost or vice versa. Rangeland that has been leased in the past might be purchased, thereby converting a variable grazing fee to the fixed costs of property taxes, mortgage interest, and interest on equity. Alternatively, an item of equipment, such as a tractor that the rancher owns, might instead be hired on a custom basis. This would allow the fixed costs of depreciation, insurance, and interest on investment to be exchanged for the per acre custom fee, a variable cost.

Building on our example in Table 5.1, suppose that the current $10 per steer per month grazing fee includes $1.80 for herding, salting, and maintenance of water facilities and fences, leaving $8.20 to be applied towards the landowner's property taxes and interest on investment. Next suppose that the cattle owner purchases the range that he had been leasing. This purchase converts the former $10 per steer per month variable cost to a combination of a variable cost of $1.80 per steer per month and an annual fixed cost of $328 per section, based on the original Table 5.1 optimum of 20 steers per section ($8.20 per month × 2 months × 20 steers). What effect will this conversion of a variable cost to a fixed cost have on optimum stocking rate? Now that the cattle owner also owns the range, is there an incentive to graze at a higher or lower intensity? Happily, this is one question where intuition and economic theory both lead to the same answer. Based on common sense alone, most observant range managers would correctly predict that if there is a switch from the pay-as-you-go arrangement of a variable grazing fee to the pay anyway case of rangeland ownership, there is a tendency to increase the stocking rate.[18] This prediction is confirmed by making the appropriate changes in our original

Table 5.5 Increased Optimum Stocking Rate Due to Conversion of a Variable Grazing Fee to a Fixed Land Cost

Variable input (steers)	Fixed input (acres)	Total product (TPP) (pounds steer gain)	Total revenue (TR) ($)	Marginal revenue (MR) ($)	Total fixed cost (TFC) ($)	Total variable cost (TVC) ($)	Total cost (TC) ($)	Marginal cost (MC) ($)	Net revenue (NR) ($)
5	640	400	200	0.50	428	18	446	0.036	−246
10	640	900	450	0.50	428	36	464	0.072	−14
15	640	1150	575	0.50	428	54	482	0.086	93
20	640	1360	680	0.50	428	72	500	0.180	180
25	640	1460	730	0.50	428	90	518	0.450	212
30	640	1500	750	0.50	428	108	536	−0.900	214
35	640	1480	740		428	126	554		186

Table 5.1 to form Table 5.5. As evident in Table 5.5, purchase of the formerly leased range increases fixed cost by $328 per section (which, of course, has no effect on optimum stocking rate). However, land purchase also reduces variable cost by $16.40 per steer for the 2-month season or by $82 per group of 5 additional steers. As predicted then, the conversion of variable cost to fixed cost increases the optimum stocking rate to slightly more than 30 steers per section. Based on this example and widespread empirical observations, we can conclude that the conversion of a variable cost to a fixed cost will increase the optimum intensity of production and vice versa.[19] Application of this concept in structuring the terms of private grazing leases could help achieve the range management goal of avoiding damage from overgrazing. Establishing private range grazing fees on a pay-as-you-go variable cost per animal rather than in terms of a pay anyway fixed cost per acre could provide both the leasor and leasee with a strong economic incentive to graze more conservatively. An interesting article by Whitson and Ragsdale (1976) makes a similar plea.

THE GENERAL MARGINALITY MODEL

Like the general production function model discussed in Appendix I of Chapter 4, the three versions of the general marginality model are based on smooth, continuous graphical or mathematical functions for TPP. The optimization principle of equating MC and MR is applied in each of the three approaches. The main difference between the optimization technique used in the four discrete data tables above and the general marginality model is that the general model contains no functional discontinuities and the optimum production intensity can be accurately specified rather than merely estimated. The three approaches to optimization of continuous functions (total revenue − total cost, MC = MR, and value of marginal product) will be demonstrated graphically using general algebraic units of measure along the axes rather than specific numbers.

Total Revenue Minus Total Cost Approach

Total variable cost (TVC) is plotted against output (Y) in the top graph of Figure 5.2. By definition, TVC is the product of the number of units of variable input (A) used in production times the price per unit (P_A). Due to the law of diminishing returns, TVC displays the characteristic sigmoid shape. It may be helpful in understanding TVC to relate its shape to that of the now familiar TPP. The geometric relationship between the two curves can be conveniently demonstrated by positioning the TPP on its side beneath the TVC curve as in Figure 5.2. This allows units of production (Y) to be measured along the horizontal axis, units of variable input (in terms of A/B) to be measured from top to bottom along the vertical axis in the lower graph, and units of variable input (in terms of $A \times P_A$) to be measured from bottom to top along the

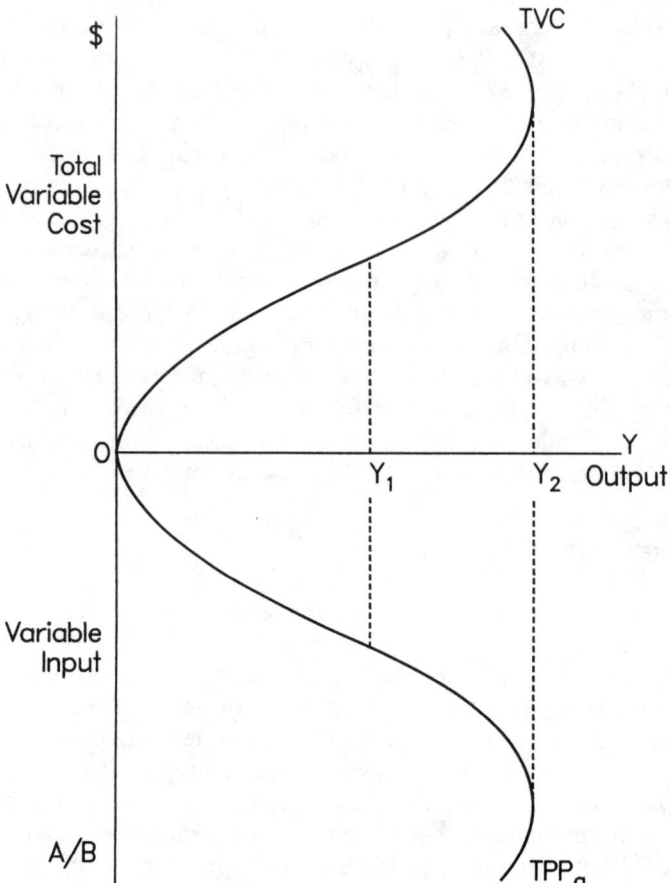

Figure 5.2 Geometric relationship between TVC and TPP curves.

vertical axis in the upper graph. Between the origin and Y_1, TVC increases at a decreasing rate (is concave) because TPP increases at an increasing rate (is convex). This may be logically interpreted to mean that since each additional unit of A is adding increasing amounts of production (Y), each additional unit of Y is adding decreasing amounts of dollar costs to TVC. At Y_1, the inflection point,[20] TVC becomes convex and TPP turns concave, meaning that since each successive unit of A used adds less to the production of Y, each successive unit of Y must add more to TVC. At Y_2, the last unit of A employed adds nothing to output (TPP has reached a maximum) and TVC bends back, indicating that beyond the peak of TPP, additional A increases TVC but suppresses production. Beyond Y_2, it actually costs more to produce less and the reader will recognize Y_2 as the output level corresponding to the boundary of production Stages II and III.[21]

We are now ready to combine the TVC curve of Figure 5.2 with a TFC curve to derive the TC needed to calculate optimum output. By definition, TFC does not vary with changes in output, causing the TFC curve in Figure 5.3 to appear as a straight horizontal line whose vertical distance denotes the dollar cost of the fixed input. As defined above, TC = TVC + TFC, which in Figure 5.3 translates to the vertical summation of the TFC and TVC curves to form the TC curve. The TVC and TC curves are identical in shape and their only difference is vertical position. At any given Y, TC lies above TVC by a vertical distance (and dollar amount) equal to TFC.

In Figure 5.4, the TC curve is combined with the total revenue (TR) curve. As previously defined, TR = $Y \times P_Y$, resulting in a straight TR line with a constant slope of $TR/Y = (Y \cdot P_Y)/Y = P_Y$. By definition, optimum output level, Y_0, is the one yielding maximum profit (TR − TC). In Figure 5.4, Y_0 corresponds to the output where the vertical distance, AB, between the TR and TC curves is greatest. This point is easily located by drawing a straight line parallel to the TR curve and tangent to the TC curve. At Y_0, the slope of the TR curve at point A, $TR/Y = (Y \times P_Y)/Y = P_Y =$ MR, is equal to the slope of the TC curve at point B, $\Delta TC/\Delta Y =$ MC. Thus, the TR − TC approach is simply an application of the concept of marginality. It will be noted that at points C and D, TR equals TC and a vertical projection of these points to the horizontal axis denotes two "break even" levels of production. While not used as often in practice as the MC = MR approach that follows, the TR − TC

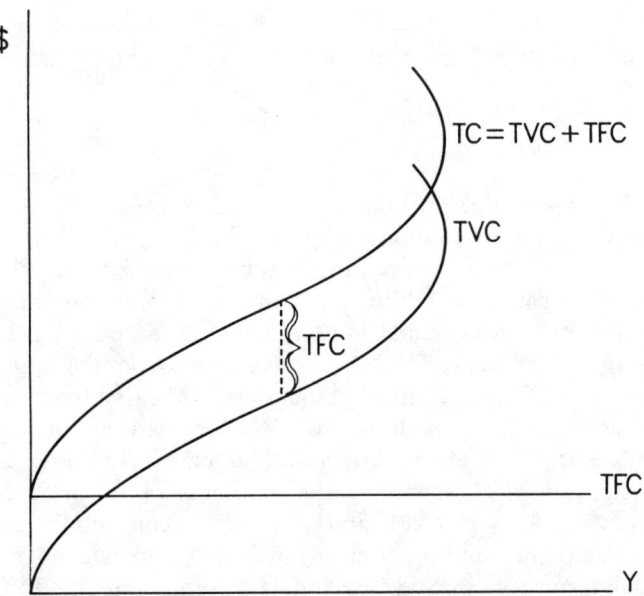

Figure 5.3 Derivation of TC by vertical summation of TVC and TFC.

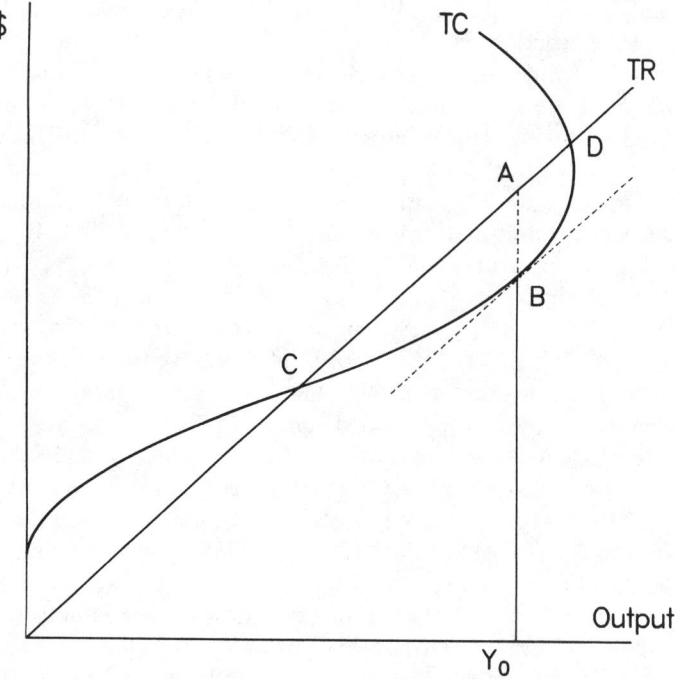

Figure 5.4 Determination of optimum output as the maximum vertical distance between TR and TC curves.

method often aids in understanding graphical solutions to the optimum intensity question.

MC = MR Approach

Employing the method discussed in Appendix I of Chapter 4, MC, AVC, and ATC curves can be derived from the continuous TVC and TC curves of Figure 5.3. The marginal and average cost curves can then be combined with the marginal revenue curve to calculate optimum output level and minimum acceptable prices. By definition, AVC equals TVC divided by output (Y). In Figure 5.5 AVC is derived as the slope, TVC/Y, of rays drawn from the origin to various points along the TVC curve. Moving along the TVC curve from the origin to point 2, the slopes of the rays drawn to TVC become ever smaller. Beyond point 2, the slope of TVC becomes larger and larger. Thus at point 2, ray R_2 just comes tangent to TVC, denoting the minimum of AVC. Since AVC = TVC/Y = ($A \times P_A$)/Y = P_A/APP and since P_A is constant in the purely competitive market, point 2 and the resulting output, Y_2, are where APP is maximized (the boundary between Stages I and II). Point 3 on the TVC curve yields the maximum output, Y_3, and, of course, corresponds to the peak of the TPP curve (the beginning of Stage III).

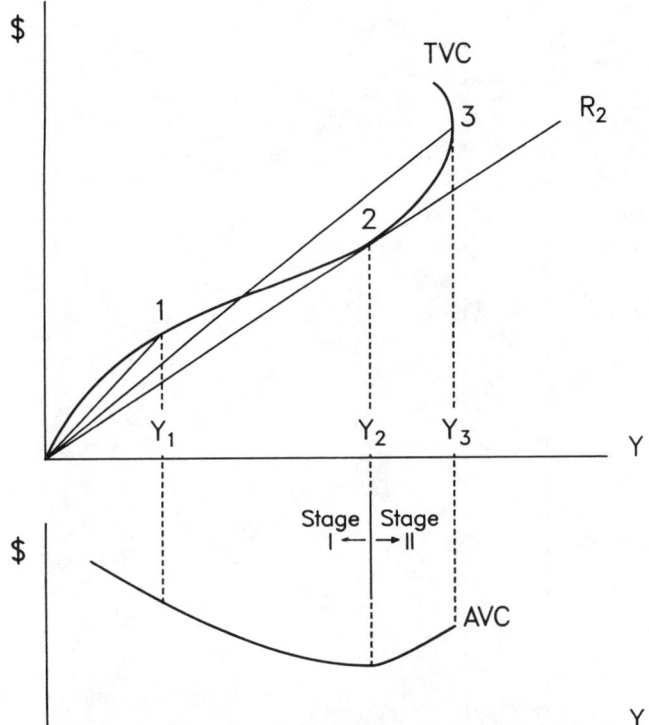

Figure 5.5 Derivation of the AVC curve from the TVC curve.

In Figure 5.6 an ATC curve is derived from the slope of rays drawn from the origin to the TC curve, as just described for AVC. The ATC minimum occurs where ray R_3 just comes tangent to TC (point 3 on the TC curve corresponding to an output of Y_3).

Finally in Figure 5.6, the MC curve is derived from either the TVC curve or the identically shaped TC curve. Beginning at the origin and moving out along the TC curve, MC, measured as $\Delta TC/\Delta Y$ or the slope of TC at a given point,[22] becomes smaller and smaller up to point 1, the inflection point[23] where TC changes from a curve that is concave to one that is convex. Beyond output Y_1, MC increases at an increasing rate until it reaches an infinite[24] value at Y_4, the output corresponding to the beginning of Stage III. As was true of the relationship between MPP and APP in Appendix I of Chapter 4, as long as MC lies below AVC, AVC decreases and wherever MC exceeds AVC, AVC must increase. Thus MC intersects AVC cost at output Y_2, the minimum of AVC. For the same reason, MC also intersects ATC at output Y_3, the minimum of the ATC curve. Figure 5.6 now contains all cost curves necessary to calculate optimum output (Y) and minimum acceptable prices.

In Figure 5.7, the MC, AVC, and ATC curves derived in Figure 5.6 are combined with an MR line, providing the most commonly used technique of

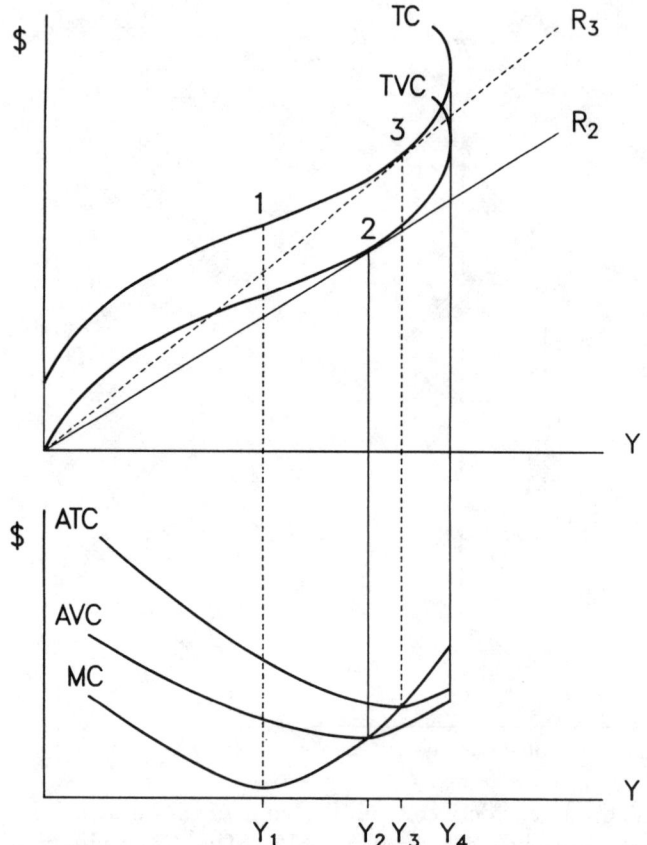

Figure 5.6 Derivation of AVC, ATC, and MC curves from TVC and TC curves.

calculating optimum output. By definition, MR = $\Delta TR/\Delta Y = \Delta Y \times P/\Delta Y = P_Y$,[25] or in terms of Figure 5.7, MR is a straight horizontal line positioned exactly P_1 dollars above the horizontal axis. Intersection of MC with MR at Y_1 marks the optimum level of output to maximize net revenue (profit). Profit, by definition, consists of TR ($Y \times P_Y$ or the rectangle OY_1CP_1) minus TC ($Y \times ATC$ or the rectangle OY_1BA). This difference between TR and TC is true economic profit, the net revenue left after all costs have been paid, including a competitive rate of interest on investment.

Suppose price drops to P_2. This would cause optimum output to decrease to Y_2, the intersection of MR and MC.[26] But what about profit? The drop in price not only cancels all economic profit formerly earned at P_1 but actually brings an economic loss. At MR_2, economic profit is negative and consists of TR − TC or the rectangle P_2DEF. Thus a loss of $F - P_2$ dollars is incurred on every unit of Y sold. How should the manager respond to this unfavorable situation? Would he be better off to continue to produce in the face of this loss or to close down the operation? A careful examination of the relationship

OPTIMUM INTENSITY OF PRODUCTION

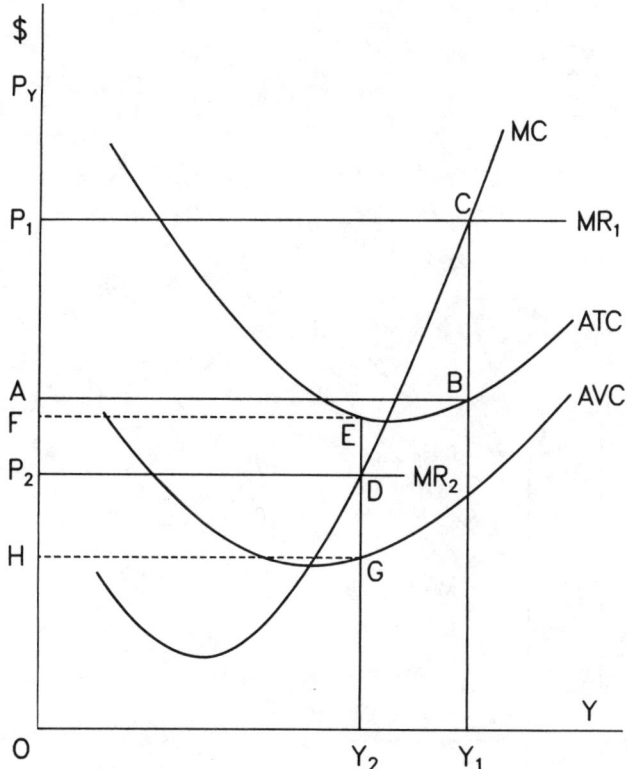

Figure 5.7 Determination of optimum output by equating MC and MR curves.

between TC and its components, TFC and TVC, will help answer this question. At price P_2 and the resulting optimum output level Y_2, TC is measured as the large rectangle OY_2EF or, by definition, ATC × Y or $OF \times Y_2$. TC can now be partitioned into its two components, TFC and TVC. TFC, by definition, equals AFC × Y and AFC is represented in Figure 5.7 as ATC − AVC. Thus at price P_2 in Figure 5.7, TFC is the quantity $GE \times Y_2$ or the rectangle $HGEF$. If the operation were completely closed down, there would be no variable costs but the fixed costs would still have to be paid resulting in a net loss exactly equal to TFC. However, the P_2 optimum output yields a TR of P_2Y_2 (the rectangle OY_2DP_2), which is sufficiently high to not only cover all variable costs (the rectangle OY_2GH) but a portion (the rectangle $HGDP_2$) of the fixed costs. Thus the net loss (the rectangle P_2DEF) produced at the optimum output Y_2 is less than the loss that would result if the operation ceased. For this reason, the optimum output Y_2, corresponding to the intersection of MC and P_2 in Figure 5.7, is called a minimum loss optimum.

Now suppose price fell to P_3 as shown in Figure 5.8. What is the optimum output level at this lower price and, even more important, should we produce at the specified optimum or should the operation cease? Again, partitioning TC

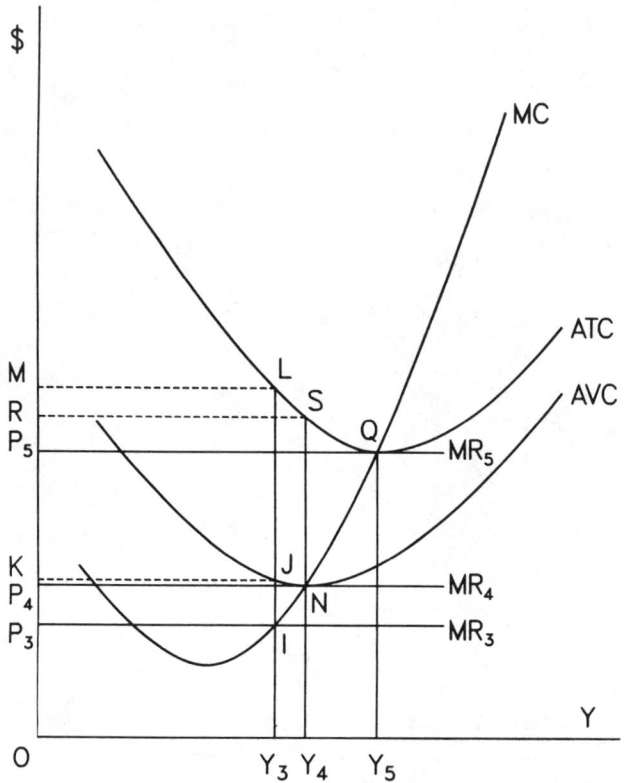

Figure 5.8 Short-run and long-run minimum acceptable prices.

into its TFC and TVC components will help guide us to the correct answers. At output Y_3 where MC intersects MR_3, TC (the rectangle OY_3LM) exceeds TR (the rectangle OY_3IP_3) by the net loss at the new optimum output, rectangle P_3ILM. If instead of producing at Y_3, all production were stopped, the short-run loss would be the smaller rectangle $KJLM$, the TFC which must be paid in the short run regardless of output level. Price P_3 is so low that not only is there no revenue to apply toward the fixed costs but only a portion (rectangle OY_3IP_3) of TVC (rectangle OY_3JK) is covered. Production at the P_3 optimum yields an extra loss (rectangle P_3IJK) over and above the short-run TFC loss that would result if the operation were shut down. Thus price P_3 is unacceptably low and a smaller net loss would be earned by going out of business than by continuing the operation.

Minimum Acceptable Prices Careful study of Figure 5.8 reveals that the minimum acceptable short-run price is one equal to the minimum point on the AVC curve. Production at any price below this point yields a loss in excess of TFC and any price above this point covers TVC as well as a portion of TFC.

OPTIMUM INTENSITY OF PRODUCTION

At price P_4, which is exactly equal to the minimum point on the AVC curve of Figure 5.8, the manager would be indifferent between producing or not producing. Either way the net loss incurred will equal the short-run fixed costs (rectangle P_4NSR).

By definition, all inputs are variable in the long run. Thus the minimum acceptable long-run price is P_5, which is equal to the minimum point on the ATC curve of Figure 5.8. At P_5 all costs of production (including a competitive rate of interest on investment) are covered. If faced with a long-run price below P_5 the rational manager would liquidate his holdings and move his capital into an alternative investment that did promise to completely cover long-run production costs. If price exceeded P_5, a pure or economic profit would be earned since total revenue would exceed total costs and, by definition, total costs already include a competitive rate of return on investment. A price higher than P_5 is a short-run phenomenon since the expectation of rates of return in excess of those earned in alternative investments would bring a flood of capital into the industry in question, increasing supply, and driving price down until it again equalled the minimum of the ATC curve.

The concept of minimum acceptable short-run price is crucial to the principle of supply discussed in Chapter 3 above. The supply curve representing the individual firm's (say the individual cow–calf operator's) willingness to sell is defined as the portion of the firm's MC curve above the minimum of AVC. At any price below the minimum acceptable price the firm ceases production and at any price equal to or above minimum AVC the firm produces where MC equals price. The supply curve (S) for the entire industry, then, is simply the horizontal summation of the portions of the MC curves of all individual firms lying above minimum AVC. Figure 5.9 demonstrates the simple case of a feeder calf industry composed of only two individual cow–calf ranches. In this simple example, industry supply (S) of feeder calves is calculated by summing the number of calves offered for sale by the two individual ranches at each calf price.

Figure 5.10 demonstrates the interrelationships between the industry supply curve for feeder calves, the industry demand curve, the industry equilibrium price, and the demand curve faced by the individual ranch. The industry demand curve for feeder calves is the horizontal summation of the demand curves (or VMP curves as described later in this chapter) of all individual firms purchasing feeder calves. The equilibrium feeder calf price is formed by the intersection of the industry supply and demand curves. The demand curve faced by the individual rancher is a straight horizontal line (the equilibrium price or MR line) and is said to be "perfectly elastic" since the individual rancher can sell all the calves he can produce at the equilibrium price. Because of his extremely small contribution to total supply of feeder calves, he cannot influence either the industry supply curve or industry equilibrium price. If he attempted to charge a price in excess of the market equilibrium, his calves would have no buyers and, of course, he would not

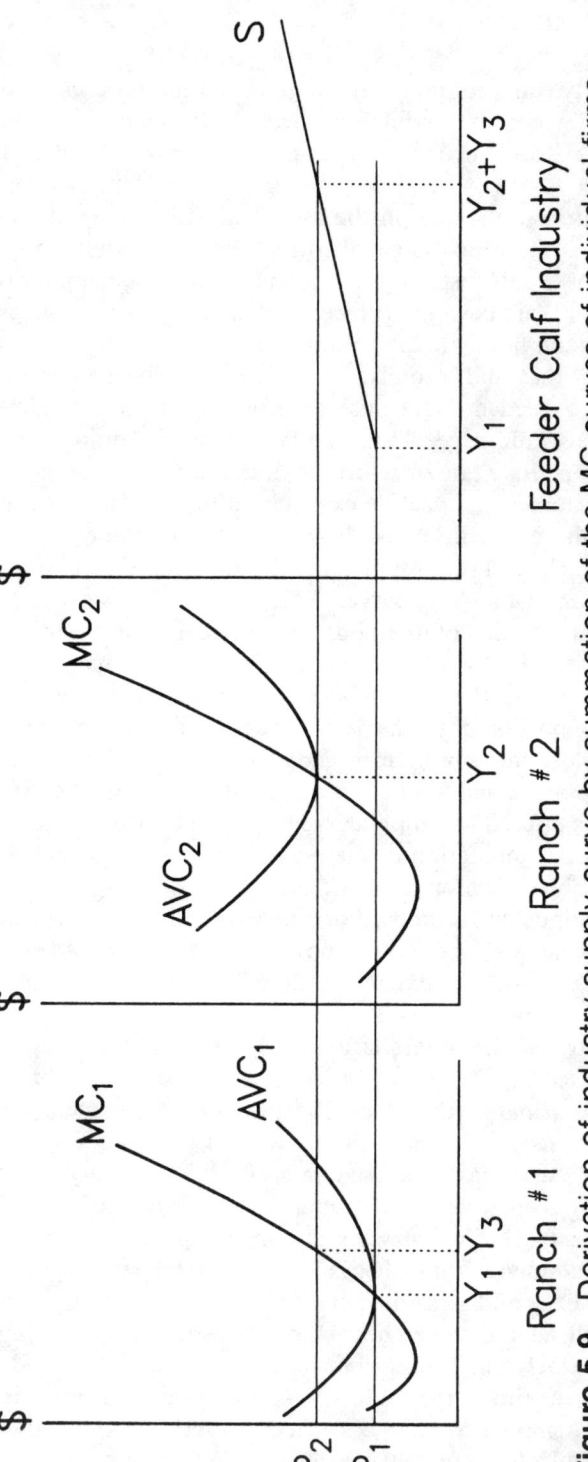

Figure 5.9 Derivation of industry supply curve by summation of the MC curves of individual firms.

OPTIMUM INTENSITY OF PRODUCTION

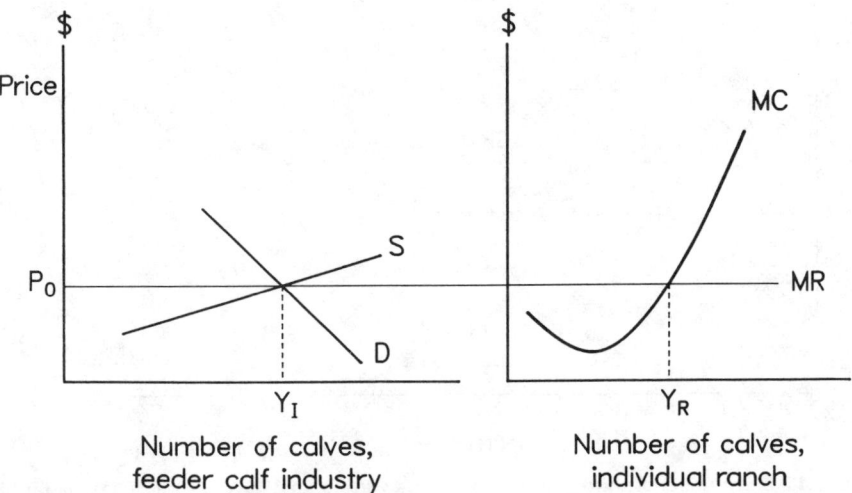

Figure 5.10 Interrelationship between the industry supply curve, industry demand curve, equilibrium price, and the demand curve faced by the individual feeder calf producer.

accept a price lower than the market equilibrium. Because of his noninfluential position due to his small share in the total market, the individual producer is said to be a "price-taker."[27]

The Long Run The long run, by definition, comprises a period of sufficient length that all inputs are variable. The long run allows enough time for new managers to be hired and old ones fired, for the most durable machines and improvements to be completely amortized, and, perhaps most important, for substantial adjustments to be made to the size or scale of the operation. Firm size has been defined in terms of resources that are fixed in the short run (Faris and Armstrong, 1963), where the short run represents a length of time in which at least one factor of production cannot be varied. Thus firm size cannot be adjusted in the short run but quantity of output can vary within the limits set by the fixed resources associated with a particular size.

As explained above, short run average cost (SRAC) curves show the effect on unit production cost of applying more or less of the variable factors to a set amount of fixed factors (Figure 5.11). SRAC curves owe their characteristic U shape to the combined effects of the spreading of fixed costs and the law of diminishing returns discussed in Chapter 4. SRAC curves first decrease as fixed costs are spread over more units of output and as production efficiency improves due to intensification of variable factors on fixed factors of production. Eventually the law of diminishing returns outweighs the continued spreading of fixed costs and the SRAC curves turn up.

A unique SRAC curve exists for each unique combination of fixed factors (each complement of machinery, buildings, and equipment). Thus SRAC curves exist that lie above those in Figure 5.11 and those shown represent the

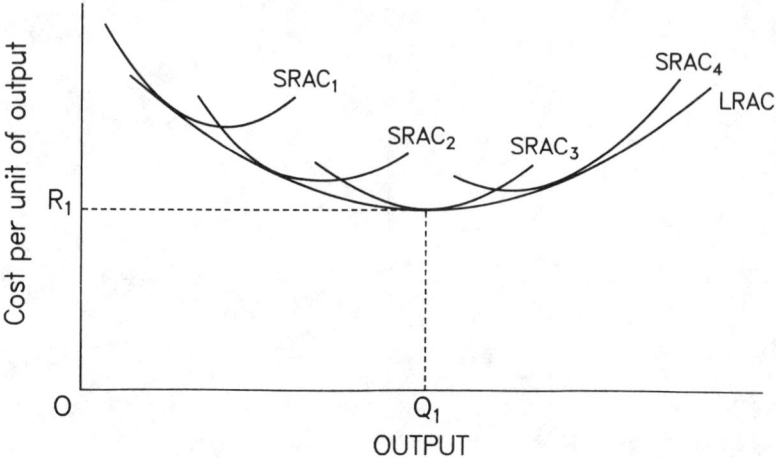

Figure 5.11 Hypothetical short-run and long-run average cost curves for firms of different sizes.

most efficienct combination of factors that is capable of producing each level of output.

The long-run average cost (LRAC) curve is formed by drawing an "envelope" curve tangent to the various SRAC curves (Figure 5.11). Each point on the LRAC curve represents a unique combination of buildings, machinery, equipment, land, and labor. As pointed out by Faris and Armstrong (1963), the LRAC curve shows the least-cost combination of inputs required to produce each given level of output. In the long run, with pure competition, firm size will tend toward the output corresponding to the minimum point on the LRAC curve (Q_1). At this point SRAC = short-run marginal cost = LRAC = long-run marginal cost = marginal revenue (= average revenue = price = R_1 in Figure 5.11), which defines equilibrium in the purely competitive market.

In a classical treatment of the theory of economies of scale (phenomena that cause unit costs to decrease as firm size increases), Viner (1950) observed that economies may be either internal (due to expansion of the individual firm) or external (due to expansion of the entire industry). Internal economies may be pecuniary (such as volume buying discounts) or technological (such as labor specialization). External economies may also take the form of either pecuniary economies (such as increased demand for raw materials allowing suppliers to expand output and ultimately pass on a lower price to buyers of raw materials) or technological (such as industry growth making it possible to increase efficiency through worker training programs).

In a discussion of cost diseconomies (phenomena that cause unit costs to increase as firms increase beyond a certain size), Heady (1952) distinguished between internal pecuniary diseconomies (such as the growth in firm size making it necessary for the firm to pay higher prices to bid labor away from other uses) and internal technological diseconomies (such as managerial ability becoming limiting as firm size increases). Distinction was also made between external pecuniary diseconomies (such as the industry expanding to the point wher inputs must be bid away from competing industries) and external technological diseconomies (such as irrigation pumping costs increasing due to the lowering of the water table as the number of farms increases).

Heady also distinguished between scale and proportionality avenues of firm size increase. Scale relationships are those in which all inputs are increased in the same proportion. If product increases by the same, greater, or smaller proportion then returns to scale are termed constant, increasing, or decreasing, respectively. A proportionality relationship is one in which at least one input is intensified upon by an increase in one or more other inputs. Thus firm size can be increased by an increase in intensity of variable input used in combination with the fixed factor or an increase in the fixed input itself.

Internal technological economies due to proportionality adjustments (increased capital : labor ratio) were deemed the most important of all farm cost economies by Heady (1952). Heady also observed that capital restrictions may prohibit the individual firm from using the LRAC curve as a planning curve. Also, the LRAC curve is only an approximation of the achievable planning curve because of the indivisibility of machinery and other inputs. Once the set of machinery (fixed factor) has been selected using the LRAC as a planning curve, the LRAC curve becomes irrelevant and the operator is restricted to the particular SRAC curve corresponding to the machinery combination selected. Empirical LRAC curves for Arizona and Utah cattle ranches have been estimated by Martin and Goss (1963) and Workman and Hooper (1971).

The Value of Marginal Product (VMP) Approach

Optimum Input Level The third marginality technique to be discussed in this chapter, the value of marginal product (VMP) approach, specifies the maximum profit optimum in terms of the quantity of input to be used rather than the quantity of output to be produced. This approach will be illustrated using the now familiar steer example from Table 5.1. The original hypothetical data of Table 5.1 appear in a slightly modified form in Table 5.6. Note that the total value product (TVP) column, as the name implies, is calculated as the product of TPP $\times P_Y$ and is identical to the total revenue (TR) column of Table 5.1. The value marginal product (VMP) column, while analogous to the marginal revenue column of Table 5.1, is very different in one important respect. VMP is calculated as the marginal income with respect to additional variable input (steers) rather than with respect to additional product (pounds). Thus at a stocking rate of 10 steers per section, VMP = $\Delta \text{TVP}_A/\Delta A$ = (\$450 − \$200)/(10 − 5) = \$50. Similarly, marginal factor cost (MFC) is calculated as the additional cost with respect to additional variable input rather than output. At 10 steers per section, then, MFC = $\Delta \text{TVC}/\Delta A$ = (\$200 − 100)/(10 − 5) = \$20. MFC remains constant at \$20 at each stocking rate since a \$10

Table 5.6 Determination of the Optimum Number of Steers by the VMP Method

Variable input (steers)	Total product (TPP) (pounds steer gain)	Total value product (TVP) (TPP × P_Y) ($)	Value marginal product (VMP) ($)	Total variable cost (TVC) ($)	Marginal factor cost (MFC) ($)	Total fixed cost ($)	Net revenue ($)
5	400	200		100		100	0
			50		20		
10	900	450		200		100	150
			25		20		
15	1150	575		300		100	175
			21		20		
20	1360	680		400		100	180
			10		20		
25	1460	730		500		100	130
			4		20		
30	1500	750		600		100	50
			−2		20		
35	1480	740		700		100	−60

monthly grazing fee is associated with each additional steer run for the 2-month grazing season. The optimum level of variable input occurs where VMP equals MFC or slightly more than 20 steers per section, the same optimum specified above by the two other marginality approaches. Inspection of the net revenue column confirms 20 steers per section as the optimum stocking rate since all other stocking rates yield less than $180 net revenue.

Figure 5.12 graphically displays the data of Table 5.6. The TVP curve is identical to the TPP curve of Figure 5.1 except that steer gain is expressed in terms of money rather than weight. Similarly, VMP is the monetary counterpart of the MPP curve discussed in Chapter 4. Just as in Table 5.6, equating VMP and MFC leads to an optimum level of variable input of slightly more than 20 steers per section. VMP actually represents the value of the gain produced by the last unit of variable input while MFC is the cost attributable to the last unit of variable input. Thus equating VMP and MFC represents one more application of the MC = MR rule for maximizing net return.[28]

It is sometimes helpful in understanding the VMP approach to visualize the VMP curve as the firm's demand curve for variable input. Thus VMP in Figure 5.12 represents the amount that each additional steer is expected to contribute to ranch income (that is, what one more steer is worth to the ranch as a producing input). In contrast, the MFC curve can be thought of as the variable input supply curve facing the firm. Thus MFC represents what each additional steer costs the ranch.[29] For detailed examples of the VMP approach involving mathematical functions, see Workman and Quigley (1974) and Workman and McCormick (1977).

OPTIMUM INTENSITY OF PRODUCTION

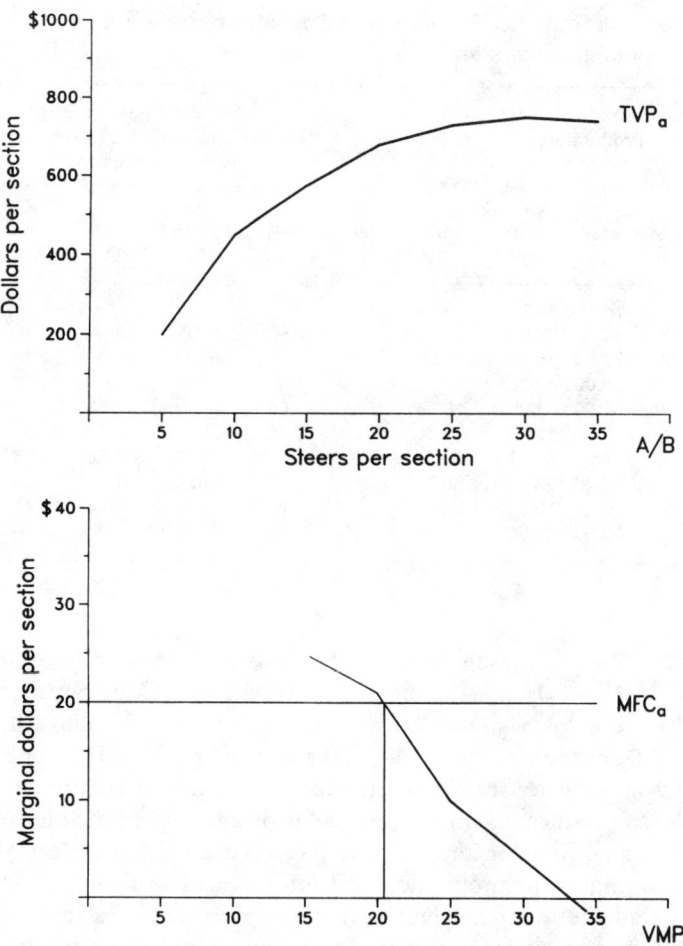

Figure 5.12 Derivation of the VMP curve from TVP and determination of optimum input by equating VMP and MFC.

Optimum Allocation among Several Enterprises Perhaps the most useful application of the VMP approach is in the allocation of a limited amount of variable input between two or more potential uses. Suppose we have the opportunity to purchase up to a total of 4000 acre feet of irrigation water at $15 per acre foot and that the water is to be allocated between two 1000-acre native hay meadows, the first a loamy, well drained site and the second a poorly drained clay site. Biological and monetary production responses to various irrigation rates on each of two sites are shown in Table 5.7. Our irrigation decision problem consists of two equally important questions: For maximum net return (1) how many total acre feet should be purchased and (2) how should the total amount be allocated between the two distinct meadow sites? These questions are easily answered by equating the return produced by

Table 5.7 Hypothetical Production, Returns, and Costs from Irrigation Water Application to Two Meadow Sites

Acre feet irrigation water applied per acre	Loam site Per acre hay production, revenue, and costs				TVC $15 / acre foot ($)	MFC ($)	Clay site Per acre hay production, revenue, and costs			
	TPP (tons)	TVP $40/ton ($)	VMP ($)	Net return ($)			TPP (tons)	TVP $40/ton ($)	VMP ($)	Net return ($)
0.25	0.35	14.00		10.25	3.75		0.20	8.00		4.25
0.50	0.75	30.00	64.00	22.50	7.50	15	0.60	24.00	64.00	16.50
0.75	1.20	48.00	72.00	36.75	11.25	15	0.90	36.00	48.00	24.75
1.00	1.60	64.00	64.00	49.00	15.00	15	1.00	40.00	16.00	25.00
1.25	1.80	72.00	32.00	53.25	18.75	15	1.08	43.20	12.80	24.45
1.50	1.90	76.00	16.00	53.50	22.50	15	1.15	46.00	11.20	23.50
1.75	1.96	78.40	9.60	52.15	26.25	15	1.21	48.40	9.60	22.15
2.00	2.00	80.00	6.40	50.00	30.00	15	1.22	48.80	1.60	18.80
2.25	2.00	80.00	0	46.25	33.75	15	1.20	48.00	−3.20	14.25
2.50	1.95	78.00	−8.00	40.50	37.50	15	1.18	47.00	−4.00	9.50

an acre foot of water (VMP) on each site with the price per acre foot (MFC). Inspection of Table 5.7 reveals that at the $15 price, 1.5 acre foot per acre should be applied on the loam site (1500 acre feet total) and the clay site should receive only 1.0 acre feet per acre (or 1000 acre feet in all).[30] Thus only 2500-acre feet of irrigation water should be purchased from the 4000 available. While we would like to produce maximum hay yields of 2 tons per acre on the loam site and 1.22 tons per acre on the clay site (both requiring 2 acre feet of water per acre), according to the production and price figures of Table 5.7 we cannot afford to use all the water available. At the application rates selected the VMP values for the two sites are equal to MFC and equal to each other and we may be assured that we have found both the optimum quantity of water to apply and the optimum allocation of that quantity between the two sites. The reader should convince himself that any change in either water quantity or allocation in Table 5.7 would reduce net return.

It is often helpful to graph the VMP and MFC information to facilitate decision-making when changes occur in either the price of the variable input (MFC) or the quantity available for purchase and use. Figure 5.13 allows decisions regarding optimum amounts and allocation of irrigation water to be made for any possible water price or availability situation. For example, if water costs $15 per acre foot ($MFC_1$), slightly more than 1.5 acre feet should be applied to each acre of loam site and the clay site should be irrigated at a rate of just over 1 acre foot per acre (a total of about 2500 acre feet to be applied to the 2000 acres just as in Table 5.7 above).

Now suppose water price increased from $15 to $48 per acre foot. How much water should we buy now and how should it be allocated between the two meadows? As shown in Figure 5.13, both these question are easily answered by equating the $48 price (MFC$_2$) with the two original VMP curves. At the higher price, application rates should be decreased, of course, but due to its flatter VMP curve (indicating the greater capacity of the loam site to productively absorb the application of irrigation water), application on the loam site should be cut back more than on the clay site. Irrigation on the loam site should be reduced by just over 0.375 acre feet while a decrease of less than 0.35 acre feet is required on the clay site. Thus a total of only 1875 of the 4000 acre feet available should be purchased at the new water price.

Next suppose that water price returns to the original $15 per acre foot (MFC) and things are just as they were except that only 1875 acre feet are available for purchase instead of the original 4000 acre feet. Since at the $15 price it was previously profitable to use 2500 acre feet we know immediately that we should purchase all 1875 acre feet available. But how much water should be applied to each site? At the $15 price we would like[31] to apply 1.5 acre feet to the loam site and 1.0 acre feet to the clay site (a total of 2500 acre feet just as before) but there simply is not enough water. Our problem is actually one of allocating the available 1875 acre feet between the two sites so as to maximize total net return. Of course we could simply irrigate the clay site at its optimum rate as dictated by the $15 price (1.0 acre feet) and apply the remaining water (875 acre feet or 0.875 acre feet per acre) to the loam site. Referring to Table 5.7, this would let us produce a per acre net return of $25 on the 1000-acre clay site and somewhere between $36.75 and $49 per acre on the loam meadow for a combined total net return of at least $61,750. Alternatively, we could apply the 1.5-acre feet optimum rate to the loam site (producing a net return of $53,500) and apply the 375-acre feet remainder to the clay site (producing a net return of between $4250 and $16,500) for a combined total net return of at least $57,750. Or we might "back off" one increment from the loam site optimum, applying 1250 acre feet on the loam site and the remaining 625-acre feet balance to the clay site giving a combined total net return of at least $69,750. This would improve our net return position since the amount gained ($16.50 − 4.25 = $12.25 per acre) by increasing the clay site application rate by 0.25 acre feet greatly exceeds the net return lost ($53.50 − $53.25 = $0.25) by decreasing irrigation on the loam site. This trial and error procedure would eventually lead us to an optimum solution. But how can we more directly determine the optimum allocation of the limited water between the two sites?

Previously we equated MFC with the two VMP curves to determine the optimum allocation. But the $15 price (MFC) is now irrelevant due to the 1875-acre feet limitation. However, according to the rule of marginality we still want the last 0.25-acre foot increment of irrigation water applied to each of the two sites to produce an equal amount of revenue. To find our elusive optimum, then, we need only to slide the MFC$_1$ line up along the two VMP curves in Figure 5.13, carefully keeping it horizontal (a ruler works well), until we find the two application rates that sum to 1.875 acre feet per acre. The horizontal

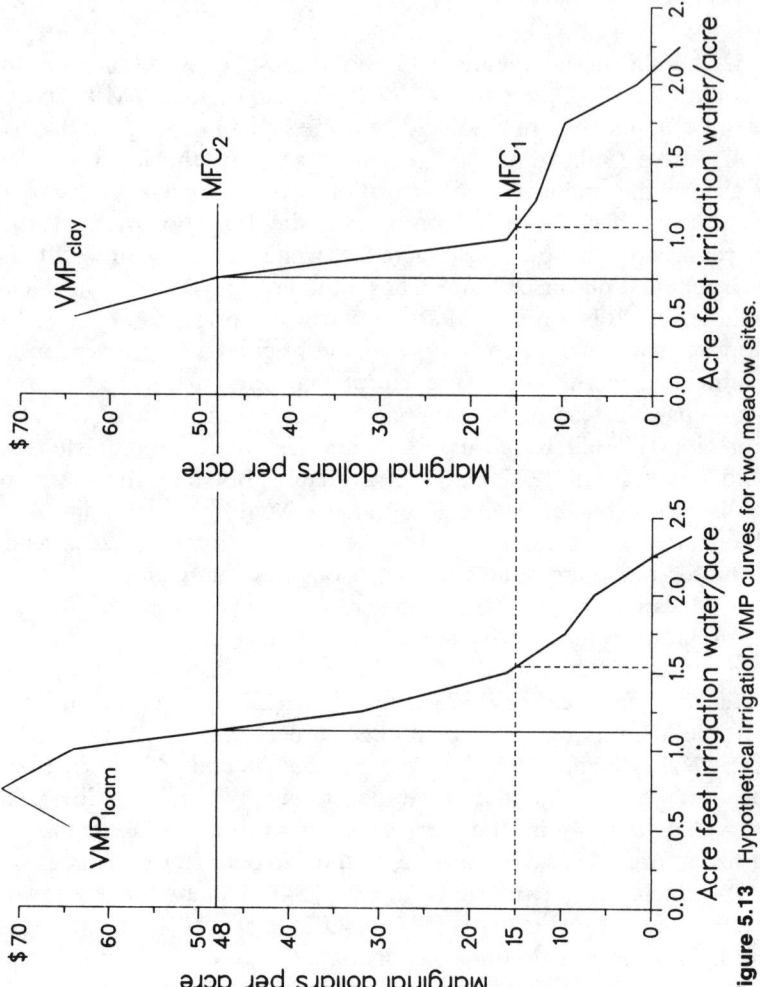

Figure 5.13 Hypothetical irrigation VMP curves for two meadow sites.

line producing this particular allocation is, by coincidence, identical to MFC_2 and the resulting optimum irrigation rates are 1.125 acre feet and 0.75 acre feet on the loam and clay sites, respectively. This particular allocation guarantees maximum combined net return from the two sites (something over $73,750 as shown in Table 5.7) and just exactly exhausts the available irrigation water.

Inspection of Figure 5.13 reveals that regardless of the price of irrigation water (MFC), the optimum allocation rate is higher on the loam site than on the clay site. This illustrates an important general rule of applying limited variable inputs. Whether irrigation water or range improvement practices such as herbicide, fertilizer, or prescribed burning, variable inputs should be applied to the "best" (most productive) sites first.

In this section we have dealt with the relatively simple case of competition for variable resources between only two uses. The VMP approach works well for algebraic solution for allocation problems involving numerous enterprises or variable input uses. The solution is determined simply by following the marginality rules which state that (1) value produced by the last increment of variable input used (VMP) must be equal for each enterprise, and (2) for the case of unlimited capital (or other unlimited resources), that the VMPs for each enterprise also equal the cost of the last increment of variable input used (MFC).[32]

LITERATURE CITED

Bement, R. E. 1969. A stocking-rate guide for beef production in bluegrama range. *J. Range Manage.* 22:83–86.

Dean, G. W. and H. O. Carter. 1960. Cost-size relationships for cash crop farms in Yolo County, California. California Agricultural Experiment Station, Giannini Foundation Research Report No. 238. 63 pp.

Faris, J. E. and D. L. Armstrong. 1963. Economies associated with size, Kern County cash-crop farms. California Agricultural Experiment Station, Giannini Foundation Research Report No. 269. 121 pp.

Gray, J. R. 1968. *Ranch Economics*. Iowa State University Press, Ames, IA. Pp. 116–120.

Heady, E. O. 1952. *Economics of Agricultural Production and Resource Use*. Prentice-Hall, Englewood Cliffs, NJ. 850 pp.

Heady, H. F. 1975. *Rangeland Management*. 1st Ed. McGraw-Hill, New York. Pp. 119–121.

Hildreth, R. J. and M. E. Riewe. 1963. Grazing production curves. II. Determining the economic optimum stocking rate. *Agron. J.* 55:370–372.

Hopkin, J. A. 1957. Economic criteria for problem solutions in research relative to the use and development of range resources. *In Economic Research in the Use and Development of Range Resources. A Methodological Anthology*. Western Agricultural Economics Research Council Report No. 1, Berkeley, CA. Pp. 35–43.

Martin, W. D. and W. K. Goss. 1963. Cost-size relationships for southwestern Arizona cattle ranches. Arizona Agricultural Experiment Station Technical Bulletin 155. 38 pp.

Mott, G. O. 1960. Grazing pressure and measurement of pasture production. In *Proceedings of the 8th International Grassland Congress*. International Grassland Congress committee. Pp. 606–611.

Pearson, H. A. 1973. Calculating grazing intensity for maximum profit on ponderosa pine range in northern Arizona. *J. Range Manage.* 26:277–279.

Roberts, C. R. 1979. Some common causes of failure of tropical legume/grass pastures on commercial farms and suggested remedies. In Sanchez, A. P. and L. E. Tergas (Eds.) *Pasture Production in Acid Soils of the Tropics*. Centro Internacional de Agricultura Tropical, Cali, Colombia. Pp. 399–416.

Schoop, M. C. and E. H. McIlvain. 1971. Why some cattlemen overgraze—and some don't. *J. Range Manage.* 24:252.

Stoddart, L. A. 1960. Determining correct stocking rate on rangeland. *J. Range Manage.* 13:251–255.

Stoddart, L. A., A. D. Smith, and T. W. Box. 1975. *Range Management*. 3rd ed. McGraw-Hill, New York. Pp. 272–276.

Viner, J. 1950. Cost curves and supply curves. In Clemense, R. V. (Ed.) *Readings in Economic Analysis*, Vol. 2. Addison-Wesley, Cambridge, MA. Pp. 8–34.

Whitson, R. E. and B. J. Ragsdale. 1976. Variable rangeland lease considerations. *Rangeman's J.* 3:143–145.

Workman, J. P. and J. F. Hooper. 1971. Cost–size relationships of Utah cattle ranches. *J. Range Manage.* 24:462–465.

Workman, J. P. and J. R. Lacey. 1982. Base ranching decisions on sound economics. *Utah Farmer-Stockman* 102:6–8.

Workman, J. P. and P. W. McCormick. 1977. Economics of carry-over response to nitrogen fertilization of rangelands. *J. Range Manage.* 30:324–327.

Workman, J. P. and T. M. Quigley. 1974. Economics of fertilizer application on range and meadow sites in Utah. *J. Range Manage.* 27:390–393.

FURTHER READING

Blackburn, A. G., M. V. Frew, and P. D. Mullaney. 1973. Estimating optimum economic stocking rates for wethers. *J. Austral. Inst. Agr. Sci.* 39:13–23.

Bromley, D. W. 1972. A dynamic economic model of pasture and range investments: Comment. *Am. J. Agr. Econ.* 54:131.

Burt, O. R. 1971. A dynamic economic model of pasture and range investments. *Am. J. Agr. Econ.* 53:197–205.

Burt, O. R. 1972. A synamic economic model of pasture and range investments: Reply. *Am. J. Agr. Econ.* 54:131–132.

Burt, O. R. 1972. More sophisticated tools for less important problems: The history of range improvements research: Reply. *Am. J. Agr. Econ.* 54:134–135.

Caton, D. D. 1959. Selection of optimum season and intensity of grazing. In *Economic Research in the Use and Development of Range Resources. Economics of Range and Multiple Land Use*. Western Agricultural Economics Research Council Report No. 2. Davis, CA. Pp. 41–59.

Caton, D. D. and C. Beringer. 1960. *Costs and Benefits from Reseeding Sagebrush Land in Idaho*. Idaho Agricultural Experiment Station Bulletin No. 326.

Cotner, M. L. 1963. Optimum timing of long-term resource improvements. *J. Farm Econ.* 45:732–748.

Cowlishaw, S. J. 1969. The carrying capacity of pastures. *J. Br. Grassland Soc.* 24:207–214.

Doll, J. P. 1972. A comparison of annual versus average optima for fertilizer experiments. *Am. J. Agr. Econ.* 54:226–233.

Harlan, J. R. 1958. Generalized curves for gain per head and gain per acre in rates of grazing studies. *J. Range Manage.* 11:140–147.

Hooper, J. F. and H. F. Heady. 1970. An economic analysis of optimum rates of grazing in the California annual-type grassland. *J. Range Manage.* 23:307–311.

Hopkin, J. A. 1959. Economic versus ecological criteria for deciding season and intensity of grazing—a discussion. In *Economic Research in the Use and Development of Range Resources. Economics of Range and Multiple Land Use*. Western Agricultural Economics Research Council Report No. 2, Davis. CA. Pp. 73–77.

Hull, A. C., Jr. and G. J. Klomp. 1974. *Yield of Crested Wheatgrass Under Four Densities of Big Sagebrush in Southern Idaho*. USDA Technical Bulletin No. 1483. 38 pp.

Jameson, D. A. 1971. Optimum stand selection for juniper control on southwestern woodland ranges. *J. Range Manage.* 24:94–99.

Johnson, W. M. 1953. Effect of grazing intensity upon vegetation and cattle gains on ponderosa pine–bunchgrass ranges of the Front Range of Colorado. USDA Circular No. 929.

Martin, W. E. 1972. More sophisticated tools for less important problems: The history of range improvements research: A comment. *Am. J. Agr. Econ.* 54:132–134.

McCorkle, C. O., Jr. and D. D. Caton. 1962. *Economic Analysis of Range Improvement*. California Agricultural Experiment Station, Giannini Foundation Research Report No. 255. 79 pp.

Pearson, H. A. and L. B. Whitaker. 1973. Returns from southern forest grazing. *J. Range Manage.* 26:85.

Peterson, R. J., H. L. Lucas and G. O. Mott. 1965. Relationship between rate of stocking and per animal and per acre performance on pasture. *Agron. J.* 57:27–30.

Ramsbacher, H. H. 1958. The economic consequences of varying the rate of grazing on eastern Montana rangeland. M.S. thesis, Montana State College, Bozeman. 89 pp.

Riewe, M. E. 1961. Use of the relationship of stocking rate on gain of cattle in an experimental design for grazing trials. *Agron. J.* 53:309–313.

Riewe, M. E., J. C. Smith, J. H. Jones, and E. C. Holt. 1963. Grazing production curves. I. Comparison of steer gains on Gulf ryegrass and tall fescue. *Agron. J.* 55:367–369.

Rittenhouse, L. R. and F. A. Sneva. 1976. Expressing the competitive relationship between Wyoming big sagebrush and crested wheatgrass. *J. Range Manage.* 29:326–327.

Stauber, M. S., O. R. Burt, and F. Linse. 1975. An economic evaluation of nitrogen fertilization of grass when carry-over is significant. *Am. J. Agr. Econ.* 57:463–471.

Stevens, J. B. and E. B. Godfrey. 1972. Use rates, resource flows, and efficiency of public investment in range improvements. *Am. J. Agr. Econ.* 54:611–621.

TEXT NOTES

1 This assumes that steers can be *bought and sold* at the same price. If there were a price spread between light and heavy animals, it would have to be accounted for in our calculations.

2 Since the $0.50 is the MR associated with the increase from 400 to 900 pounds of gain, it is positioned in Table 5.1 halfway between output levels 400 and 900. The same rule should be followed when graphing marginal values derived from incremental changes in output data.
3 Mathematically, MR $= d\text{TR}/dY$ where TR $= f(Y)$ and $Y =$ output $=$ TPP. In the purely competitive market TR $= Y \cdot P_Y$, where P_Y is output price per unit and MR $= P_Y$.
4 The "short run" is defined in economics as a period of time sufficiently short that at least one factor of production such as machinery, improvements, or owned land is fixed. The "long run" is a period of sufficient length to make all inputs variable. This section of Chapter 5 deals with the operation of a steer enterprise in the short run.
5 In mathematical form, MC $= d\text{TVC}/dY = d\text{TC}/dY$ where TVC $= f(Y)$, TC $= g(Y)$, and $Y =$ output $=$ TPP.
6 Between 30 and 35 steers per section (beyond the peak of the TPP function) MC becomes negative due to the decrease in total output.
7 The goal of the individual manager may differ somewhat from this usually assumed ideal. Some managers may attempt to maximize profit subject to the constraint that risk is kept sufficiently low so that they can assure their bankers that mortgage payments can be consistently met. Other managers may attempt to maximize profit subject to the constraint that they have a week of free time to take a hunting trip or attend a national livestock show. The assumption of unlimited capital must also often be altered. Operating capital is almost always limited and clearly if the manager could earn more than 5 percent ($180 - 175$)/($400 - 300$) on his limited funds in some alternative investment, his optimum stocking rate would be something less than 20 steers per section. Although the analysis must be modified to fit the true goals and financial circumstances of the individual manager, the maximum profit–unlimited capital assumptions should be used as the starting point for optimum intensity determinations.
8 The true optimum stocking rate in Table 5.1 is slightly higher than 20 steers per section since at the 20 steer intensity MR still slightly exceeds MC. At 20 steers per section the last unit of product (actually 210 units or pounds, due to the discrete data of our example) adds $105 to revenue and only $100 to costs.
9 The all too evident "tragedy of the commons" problems associated with livestock grazing of public lands are an entirely different matter. The present discussion is confined to the private ownership case.
10 $P_Y = \$0.50$ is the price per pound of steer gain. $P_A = \$20$ is the cost of grazing one additional steer per section for 2 months ($10 per steer per month $\times$ 2 months). The price ratio is inverse with respect to MPP, which expresses the ratio $\Delta Y/\Delta A$.
11 This approach has often been referred to by my students as the "shortcut method." This method works well when the smooth continuous production function relationships estimated by computer regression analysis are not available. Under field conditions a reasonably accurate input optimum can be determined by combining an inverse price line with a production function derived by straight-line segments drawn through a few data points as in Figure 5.1. A simple small scale multiple intensity experiment will provide the necessary data and the required analytical equipment consists of a ruler and lined graph paper.
12 The specification of budget at $200 is an arbitrary choice, made in this case to

allow the correct number of pounds to be easily located on the small scale vertical axis of Figure 5.1. A budget of $1000 or even $1 would accomplish the same result.

13 At this point it should be obvious why it is per acre rather than per animal net returns that are to be maximized. The goal of the private landowner is to maximize total net revenue from his entire operation. Since land is the fixed factor of production, maximum net returns per acre is synonymous with the maximum total net returns. The manager need not concern himself with per animal performance per se. As long as (1) all costs are accounted for (as they are in Table 5.1), (2) the animals being added to (or subtracted from) a fixed amount of range are uniform in their productivity response to changes in forage availability, and (3) changes in animal condition are not sufficient to bring changes in price per pound of animals sold, specifying optimum grazing intensity in animals per unit area (20 steers per section) automatically brings optimum performance per animal (1360 pound ÷ 20 steers = 68 pounds gain per steer for the 2-month season).

14 But numerous examples of overgrazing and resultant declines in range condition exist on private lands where the profit motive would be expected to operate. Why? As mentioned above, ignorance and optimism concerning carrying capacity seem to be responsible. More specifically, the apparent (or short-run) production function lies above the actual (or long-run) production function of Figure 5.1. Thus, optimization of the apparent production function can lead to stocking rates in excess of actual maximum sustained yield.

15 Maximum sustained yield is the legitimate optimum stocking rate on some public rangelands. From the viewpoint of the individual public range permittee, the marginal cost of the administered grazing fee may be less than marginal revenue (measured in terms of savings in the costs of alternative feeds such as hay) at all conceivable stocking rates. In this situation, the stocking rate must be set according to biological criteria and, if carefully calculated, the maximum sustained yield rate is a logical target.

16 Actually at 15 steers per section, MR slightly exceeds MC and the precise optimum stocking rate is just beyond 15 steers per section.

17 Since MR still slightly exceeds MC at this stocking rate, the precise optimum is just beyond 15 steers per section.

18 In response to observations by visitors that private seasonal rangelands appear to be overgrazed, ranchers have been known to remark: "The reason I bought the land is so I would have a place to run my stock at this time of year."

19 In addition to the explanation of footnote 14, this is another reason that the long-run optimum stocking rate is lower than the short-run optimum. The long run, by definition, is sufficiently long that all inputs are variable so all short-run costs (both fixed and variable) become variable in the long run.

20 Mathematically, the inflection point occurs at the same Y in both curves. For TVC, the inflection point occurs at $d\text{TVC}/dY = \min$ where TVC $= f(Y)$. For TPP the inflection point occurs at the Y corresponding to A at $d\text{TPP}/dA = \max$ where TPP $= f(A)$.

21 It should be noted that the stages of production have previously been specified in terms of the level of input. The discussion of cost curves that follows specifies production stages in terms of output level.

22 Mathematically, MC $= d\text{TC}/dY$ where TC $= f(Y)$.

23 Mathematically, the output at which $d\text{TC}/dY = 0$, where TC $= f(Y)$.

24 Intuitively, MC must become infinite at Y_3 since additional variable input not only

25 Mathematically, MR = $d\text{TR}/dY$, where TR = $f(Y)$.

26 The analytical advantage of calculating optimum output level in terms of equating the MC and MR curves is an obvious one. No matter what changes occur in output price (or variable or fixed input prices), the direction of the appropriate change in output that the manager should make and the corresponding change in profit can be accurately predicted. Although the manager does not normally have estimates of his various cost curves (and often does not know that such curves even exist), his trial and error decisions ultimately lead him to respond to price changes as if he did and make his actions predictable. Accurate predictions of appropriate adjustments in output and resulting profit changes are based on an accurate answer to only one question: "Which curves (MR, MC, AVC, and ATC) will be shifted by the price change in question?"

27 Precisely analogous to the individual feeder calf producer as a "price taker" is the individual feeder calf buyer. The feeder calf supply curve faced by the individual buyer is a straight horizontal line (the equilibrium price line). According to this "perfectly elastic" supply curve, the small buyer can purchase all the feeder calves he wants at the equilibrium price but can obtain none at a price lower than the market equilibrium. He would not knowingly ever pay more than the market equilibrium price. Because of his weak market position, the small buyer is said to be a "price payer."

28 Mathematically, VMP_A can be written as $(\text{MPP}_A)(P_Y)$. Thus:

$$\text{VMP}_A = \text{MFC}_A,$$
$$(\text{MPP}_A)(P_Y) = \text{MFC}_A,$$
$$P_Y = \frac{\text{MFC}_A}{\text{MPP}_A} = \frac{\Delta\text{TVC}/\Delta A}{\Delta Y/\Delta A} = \frac{\Delta\text{TVC}}{\Delta Y}, \quad \text{or}$$
$$\text{MR} = \text{MC}.$$

29 The reader will recognize that what is actually being demanded and supplied in Figure 5.12 is the variable input, steer units of forage for a 2-month period. It is this variable input that produces the MPP (measured in terms of steer gain) that is ultimately sold at price P_Y. Since only the gain is being sold we are implicitly assuming that steer purchase and sale prices are the same.

30 These optimum application rates are also the optima for each site considered separately. However, as will be demonstrated below, optima resulting from correct allocation of limited variable inputs between two enterprises are often less than the optima for individual enterprises.

31 The reader will also recognize that if we were concerned with only one meadow, say the 1000-acre loam site, we would like to and could apply 1.5 acre feet per acre (a total of only 1500 acre feet even though 1875 acre feet are available). Similarly, if we were dealing with only the clay site our achievable optimum would be 1.0 acre feet per acre (only 1000 of the 1875 acre feet available). The optimum allocation that we seek above is termed a "constrained optimum" since insufficient variable input is available to achieve both of the two individual optima.

32 Mathematically,

(1) $\text{VMP}_1 = \text{VMP}_2 = \cdots = \text{VMP}_n,$
(2) $\text{VMP}_1 = \text{VMP}_2 = \cdots = \text{VMP}_n = \text{MFC}_A = P_A,$

where $1, 2, \ldots, n$ are competing uses of variable input (A).

Chapter 6

Optimum Combination of Inputs

The second basic production problem involves the question "how should the product be produced?" As with the optimum intensity question discussed in Chapter 5 above, solution of the optimum combination problem relies on the application of the law of diminishing returns and the principle of marginality.

A TABULAR PRODUCTION FUNCTION

Suppose we are involved in the production of cattle gain by feeding ten head of yearling steers hay and allowing them to graze winter range forage (Table 6.1). In order to optimize our production process so as to maximize our net return, we must first answer one of the following two questions: (1) What is the optimum mix of hay and winter range to minimize the costs of producing a given amount of steer gain? Alternatively, (2) what is the optimum mix of hay and winter range to maximize steer gain from a given operating budget?

Next, armed with the optimum mix information, we must answer a third question: (3) What is the optimum total steer gain for maximum net return (or how much, in total, should we spend on hay and winter range forage)?

Inspection of Table 6.1 reveals that hay and winter range are input substitutes. That is, steer gain can be produced exclusively with range forage or

Table 6.1 Hypothetical Monthly Gain From 10 Yearling Steers Fed Various Combinations of Winter Range Forage and Hay

AUM of winter forage (input A)	Tons of hay (input B)					
	0	2	4	6	8	10
0	0	150	250	340	420	490
5	150	340	400	450	485	520
10	250	400	490	520	540	550
15	340	450	520	540	550	555
20	420	490	540	550	555	559
25	490	520	550	555	559	560

by feeding hay in the absence of range forage. The effects of the law of diminishing returns are evident. As more and more hay is fed in the absence of winter range, total steer gain increases but at a decreasing rate. A 2-ton increase in hay from 2 to 4 tons gives a 100-pound increase in monthly steer gain but a 2-ton increase from 8 to 10 tons yields an increased steer gain of only 70 pounds. Likewise, as more range forage is grazed in the absence of hay feeding, steer gain increases at a decreasing rate. A 5-AUM (animal unit month) increase from 5 to 10 AUM yields an increased gain of 100 pounds but a 5-AUM increase from 20 to 25 AUM produces only 70 pounds additional steer gain. Steer gain can also be produced by feeding combinations of the two feeds. Increased hay and range forage in set combinations also exhibit diminishing returns. Increasing forage use from 5 AUM and 2 tons to 10 AUM and 4 tons increases steer gain by 150 pounds. A forage use increase of the same magnitude from 10 AUM and 4 tons to 15 AUM and 6 tons increases steer gain by only 50 pounds.

Table 6.1 contains only part of the information required to calculate answers to these questions. As in the optimization procedure described in Chapter 5, we must also incorporate data concerning prices of the two inputs of hay and winter range and the value of the livestock gain produced.

A GRAPHICAL EXAMPLE

For ease of analysis, let us now exhibit a portion of the production function information of Table 6.1 in graphical form. Figure 6.1 is a graphical presentation of part of the production data of Table 6.1. The kinked lines of Table 6.1 are called isoquants (equal product curves) and show the various combinations of winter range forage (input A) and hay (input B) capable of producing given quantities of steer gain. A monthly steer gain of 340 pounds (isoquant 1), for example, can be produced by feeding 6 tons of hay and no range forage, 15 AUM of winter range and no hay, 2 tons of hay combined with 5 AUM of range forage, or any combination of hay and forage lying along isoquant 1. The isoquants graphically demonstrate that winter range forage and hay are

OPTIMUM COMBINATION OF INPUTS

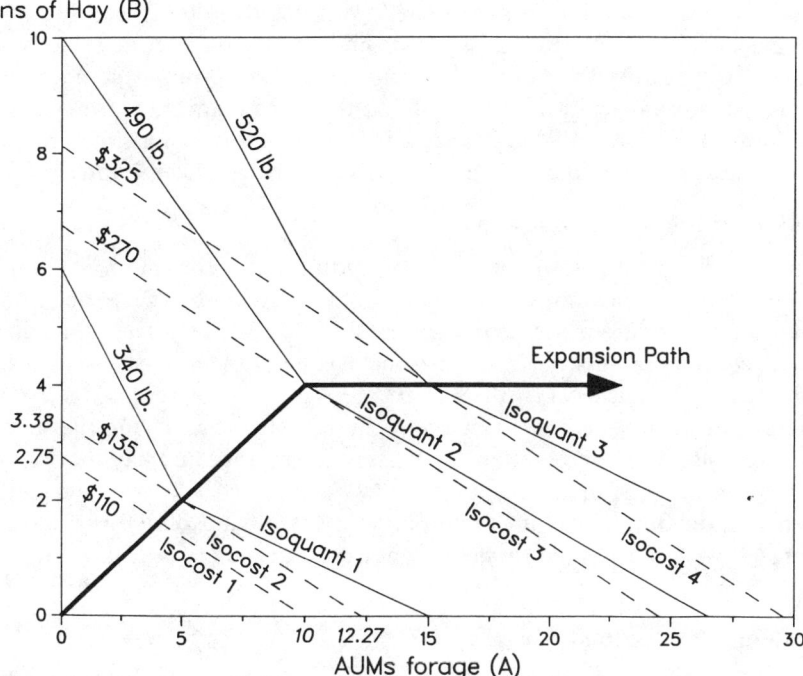

Figure 6.1 Hypothetical production functions, isocost lines, and optimum feed mixes for 10 yearling steers.

input substitutes. The quantity of hay fed can be decreased without suppressing steer gain as long as winter range forage is increased to compensate for less hay.

The Minimum Cost Combination

Let us now introduce the input price data necessary for calculating the optimum combination of winter range forage and hay to minimize the cost of producing a given amount of steer gain. Suppose that we can lease winter range for $11 per AUM and that hay is selling for $40 per ton. As with the "shortcut" method of Chapter 5, this price information is incorporated into Figure 6.1 by answering two questions. First, how much winter range can we lease with a given amount of operating capital, say $110? The answer, $110/$11 = 10 AUM is plotted on the horizontal axis of Figure 6.1 as one endpoint of isocost 1. Next, how much hay can we buy with the same $110 of capital? The answer, $110/$40 = 2.75 tons is plotted on the vertical axis as the other endpoint of isocost 1. Connecting the two endpoints with a straight line, we now have isocost (equal expenditure) 1, which shows all possible combinations of winter range and hay that can be purchased with a budget of $110. Suppose our goal is to produce 340 pounds of steer gain per month at minimum cost. In terms of Figure 6.1, we want to operate at the point on isoquant 1 where total forage costs are the lowest possible. We locate this point by moving a line parallel to isocost 1 up and to the right until it becomes tangent to isoquant 1 (the point where isocost 2 makes initial contact with isoquant 1). This tangency occurs at the kink in isoquant 1, representing a combination of 5 AUM of winter range and 2 tons of hay. Thus the minimum cost of producing 340 pounds of gain is $135 (5 AUM × $11 + 2 tons × $40).

The Maximum Product Combination

Now suppose our goal is stated in terms of finding the optimum mix of winter range and hay to maximize steer gain from a given forage expenditure. Suppose, further, that our forage budget is limited to $270. In terms of Figure 6.1, our goal is to push a straight line parallel to one of the isocost lines already plotted (isocost 1 or isocost 2) up and to the right until it represents a total expenditure of $270, our limited budget. The relevant isocost line (isocost 3) represents all possible combinations of winter range forage and hay that can be purchased for $270. Optimum solution occurs where isocost 3 comes tangent to isoquant 2, indicating that a maximum monthly steer gain of 490 pounds can be produced with our limited forage expenditure of $270 (10 AUM × $11 + 4 tons × $40).

The Optimum Production Level

We are now prepared to address the third question posed above: How much total steer gain should be produced for maximum net return? We know that the optimum production level must occur at a tangency of an isocost line with an isoquant, satisfying the requirement that each particular quantity of steer gain is produced with a minimum cost combination of range forage and hay. The line labelled "Expansion Path" in Figure 6.1 traces out the various points of tangency showing the minimum cost combinations. Thus, the optimum level of total steer gain lies somewhere on the expansion path. To determine exactly where net revenue reaches a maximum, one additional bit of information must be included, steer gain price. Suppose live weight steer price is $0.91 per pound. Applying the optimum intensity analytical techniques of Chapter 5, the data of Figure 6.1 is combined with steer price to produce Table 6.2. For example, the following Table 6.2 information is derived from the tangency of isocost 2 and isoquant 1 in Figure 6.1: winter range = 5 AUM, cost of winter range = $55 ($11 × 5), hay = 2 tons, cost of hay = $80 ($40 × 2), total feed cost = $135 ($55 + $80), total steer gain = 340 pounds, total revenue = $309 ($0.91 × 340), marginal cost = $0.40 per pound (Δ total feed cost = $135 − 0 ÷ Δ steer gain = 340 pounds − 0), marginal revenue = $0.91 per pound (steer price),[1] and net revenue = $174 ($309 − 135). After constructing Table 6.2, determining the optimum steer gain to maximize net return is simply a matter of equating marginal cost with marginal revenue. Thus, the optimum level we are seeking is slightly more than 490 pounds of steer gain (an isoquant slightly above isoquant 2). This optimum production level is verified by the corresponding maximum net revenue of $176.

Table 6.2 Hypothetical Production, Costs, and Returns From 10 Yearling Steers Fed Various Combinations of Winter Range Forage and Hay

Variable input A (winter range) (AUM)	Cost of A ($)	Variable input B (hay) (tons)	Cost of B ($)	Total feed cost (A + B) ($)	Marginal cost ($)	Total steer gain (lb)	Total revenue ($)	Marginal revenue ($)	Net revenue ($)
0	0	0	0	0		0	0		0
5	55	2	80	135	0.40	340	309	0.91	174
10	110	4	160	270	0.90	490	446	0.91	176
15	165	4	160	325	1.83	520	473	0.91	148

Most empirical examples of the use of isoquants in production decisions involve the application of two or more chemical fertilizer compounds to enhance crop production (Heady, 1952; Heady and Pesek, 1954) and few examples exist for livestock production and forage use. Notable exceptions include Brokken et al. (1976), Brokken and Bywater (1982), Burt (1978), and Heady et al. (1963).

THE GENERAL ISOQUANT–ISOCOST MODEL

In the steers, winter range, and hay example above, isoquant curves were plotted by connecting a small number of discrete data points with straight line segments. In general form, isoquants appear as smooth, continuous curves (Figure 6.2). The general isoquant–isocost model applies to any situation where two variable inputs (A and B) are being applied to one (or more) fixed inputs (C) to produce one product (Y) with either a limited or unlimited budget ($). Again, it may aid the reader's understanding to mentally substitute specific examples for general terms: range forage and hay for A and B, base property for C, and gain from a specific number of steers for Y. The general model addresses two related questions: (1) determination of the optimum mix of variable inputs A and B to minimize the cost of producing a specified amount of Y or, alternatively, determination of the optimum mix of A and B to maximize the amount of Y produced with a given expenditure for A and B and (2) determination of the optimum quantity of Y to produce for maximum net return.

A number of good textbooks clearly describe the general isoquant–isocost model. Highly recommended are those by Heady and Dillon (1961), Allen (1962), and Doll and Orazem (1978).

Derivation of the Isoquant Map

Figure 6.2 illustrates the derivation of an isoquant map (lower graph) from a three-dimensional production function (upper graph). Vertical projection of the upper three-dimensional production function (a "mountain") yields the

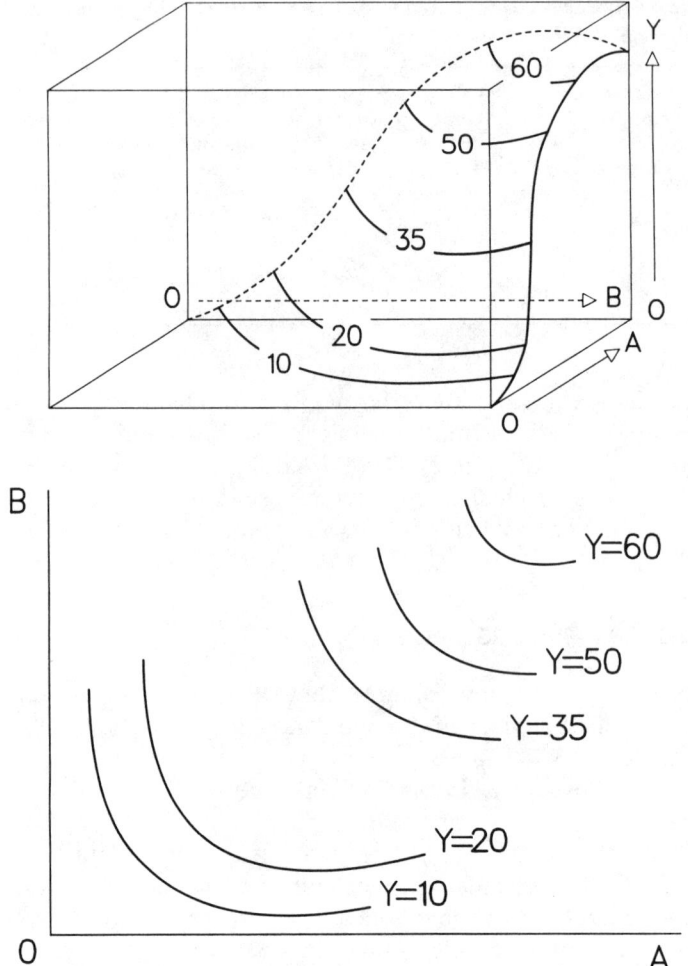

Figure 6.2 Derivation of a two-dimensional isoquant map from a three-dimensional production function.

lower two-dimensional isoquant representation (a "topog map"). Let us first examine the upper graph representing two variable inputs (A and B) capable separately or in combination of producing output Y. Beginning on one of the two horizontal axes of the upper graph, notice that production of Y (on the vertical axis) can be increased by (1) adding more units of variable input A (on the A horizontal axis), (2) adding more units of variable input B (on the B horizontal axis), or (3) adding more units of both A and B in combination. Specific levels of Y are represented by the curved "elevation lines" labelled $Y = 10$, $Y = 20$, etc. Notice, also, the obvious workings of the law of diminishing returns. Addition of A, in the absence of B, yields a sigmoid-shaped two-dimensional production function. As more units of input A are used,

OPTIMUM COMBINATION OF INPUTS   91

production of Y increases first at an increasing rate, then increases at a decreasing rate, and finally reaches a peak. Addition of B, in the absence of A, yields a similarly shaped function. And the addition of A and B, together, also produces a sigmoid-shaped function.

Now, let us direct our attention to the bottom graph of Figure 6.2. Just as a two-dimensional topog map accurately portrays the shape of a three-dimensional mountain, the isoquant map provides exactly the same information as the three-dimensional production function from which it is derived. Units of variable input A are now plotted along the horizontal axis, units of variable input B are plotted along the vertical axis, and the production level of Y is represented by the curved isoquants labelled $Y = 10$, $Y = 20$, etc.

Characteristics of Isoquants

Production economics theory describes isoquants as having four main characteristics:

1. Isoquants are "everywhere dense." In terms of the lower portion of Figure 6.2, this simply means that although isoquants are shown for only five levels of Y production, isoquants exist somewhere on the map for every possible level of Y production ($Y = 11$, $Y = 22\frac{1}{2}$, $Y = 61.4$, etc.).
2. Isoquants do not intersect. Since, by definition, a particular isoquant represents all possible combinations of A and B capable of producing a given amount of Y, isoquants cannot intersect. Such an intersection would imply that for a specific quantity of Y, say 15 units, there are two sets of combinations of A and B, each set representing all possible mixes of A and B capable of producing $15Y$. The existence of more than one such mutually inclusive set would be nonsense.
3. Isoquants slope down and to the right. The fact that isoquants are downward sloping has great practical significance. It demonstrates that the two variable inputs are input substitutes. The use of input B can be decreased while maintaining the production level of Y if the decrease in B is compensated by a simultaneous increase in input A. Likewise, Y can be maintained at a specified level and input A can be decreased if the use of input A is increased to compensate.
4. Isoquant functions are convex to the origin. Isoquants not only slope downward, but exhibit an important shape. In almost all production situations, isoquant curves are convex to the origin of the graph. This characteristic curve shape is again due to the all important law of diminishing returns. In conventional terminology, the marginal rate of technical substitution (MRTS) of input A for input B (MRTS_{AB}) decreases as we move down and to the right along the isoquant. In other words, the slope of the isoquant $\Delta B/\Delta A$, becomes less and less steep moving down along the isoquant. $\text{MRTS}_{AB} = \Delta B/\Delta A$ represents the amount of input B we can give up divided by the amount of input A we must add, while just maintaining product Y at a particular

level.[2] Thus, if Y is maintained by increasing A and decreasing B, the slope of the isoquant (MRTS_{AB}) decreases. MRTS_{AB} may also be written $\text{MPP}_A/\text{MPP}_B$.[3] Due to the law of diminishing returns, as more A is added to less B, MPP_A decreases and MPP_B increases, decreasing the ratio $\text{MPP}_A/\text{MPP}_B$.

Ridge Lines — the Relevant Stage

As mentioned above, one important isoquant characteristic is that they are downward sloping. As indicated in Figure 6.3, complete isoquant curves have a semicircular shape. However, the relevant portions of isoquant curves (the portions actually used in making input combination decisions) slope down and to the right. Isoquant Y_2, for example, is semicircular in its entirety. However, only the small, solid part of Y_2 (the EC portion) is relevant to production

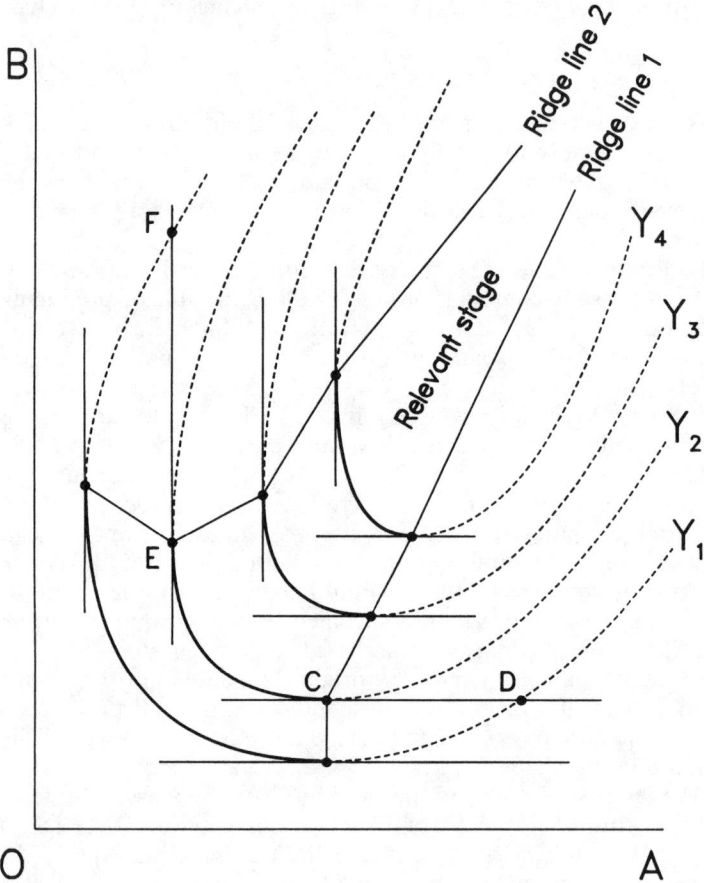

Figure 6.3 Ridge lines, relevant portions of isoquant curves, and the relevant production stage (Stage II).

decisions since production would never rationally take place on the dashed portion of Y_2. Note point C, where a horizontal line parallel to the x axis comes tangent to Y_2. Production of Y_2 would never rationally take place to the right of tangency C because it would involve a waste of inputs. The dashed portion of Y_2 slopes upward indicating that more of both inputs A and B are being used but production remains constant at Y_2. Thus, the use of additional A and B beyond point C is wasted since increased inputs (and higher costs) do not result in increased production. Note also that if sufficient additional input A is employed while holding input B constant at C, production is decreased from Y_2 to Y_1 at point D. In fact, to the right of point C, if input A is increased, input B must be increased simultaneously, just to maintain production at Y_2. Thus point C marks the lower end of the relevant portion of isoquant Y_2. Due to the law of diminishing returns, the dashed portion of Y_2 to the right of point C lies in Stage III with respect to input A. The same is true of all isoquants in Figure 6.3. Tangencies of horizontal lines with the various isoquants mark the end of their relevant portions (Stage II) and the beginning of Stage III with respect to input A. The isocline (equal slope) line connecting these tangencies (points of zero slope) on the isoquants is called a ridge line and denotes the boundary between production Stages II (the relevant stage) and III with respect to input A for the entire isoquant map. In practical terms, in order to avoid using too much of inputs A and B, production should take place only to the left or above ridge line 1.

Let us now focus on point E. For reasons similar to those described above, production would never rationally take place on the dashed portion of Y_2 above point E. Point E marks the point where a vertical line parallel to the Y axis comes tangent to Y_2. Above tangency E, Y_2 slopes upward indicating that more of both inputs A and B are being used but, again, production remains constant at Y_2. Thus, the use of more A and B beyond point E is wasted since increased inputs do not increase production. Also, if sufficient additional input B is used while holding input A constant at E, production is decreased from Y_2 to Y_1 at point F. Above point E, if input B is increased, input A must be increased simultaneously just to maintain the production level at Y_2. Thus point E marks the upper end of the relevant portion of isoquant Y_2. Again, because of the law of diminishing returns, the dashed portion of Y_2 above point E lies in Stage III with respect to input B. The same holds true for all four isoquants shown in Figure 6.3. Tangencies of vertical lines with the various isoquants mark the end of their relevant portions (Stage II) and the beginning of Stage III with respect to input B. The isocline connecting these points of equal (and infinite) slope is labelled ridge line 2 and marks the boundary between Stages II and III for the entire isoquant map. Again, in practical terms, in order to avoid using too much of inputs A and B, production should take place only below ridge line 2. Thus the relevant stage of production (Stage II) for both inputs A and B occurs on the isoquant portions lying between ridge lines 1 and 2.

The Isocost Line and the Optimum Mix

As with the preceding tabular production function example, the general isoquant–isocost model involves the calculation of answers to three questions: (1) What is the optimum input mix to minimize the costs of producing a particular amount of output? (2) What is the optimum input mix to maximize production for a particular expenditure? (3) What is the optimum output level for maximum net return? These three questions will be answered in turn.

The Minimum Cost Optimum The answer to the first question of the optimum input mix to minimize production costs begins with the isoquants of Figure 6.4. Isoquants Y_1 through Y_5 show various combinations of variable inputs A and B capable of producing five levels of output. Suppose our goal is to minimize the cost of producing Y_1 output. What we are really searching for is the lowest isocost line (smallest budget) capable of producing Y_1 output. We begin by establishing an isocost line anywhere on the isocost map and then drawing an isocost line parallel to it that is just tangent to Y_1. Suppose we begin with isocost 1, derived as described above, using an operating budget of $1. Next, a parallel isocost line (isocost 2) is drawn tangent to Y_1 and we have located the optimum mix of variable inputs (A_1 and B_1) at point C to minimize the costs of producing Y_1.

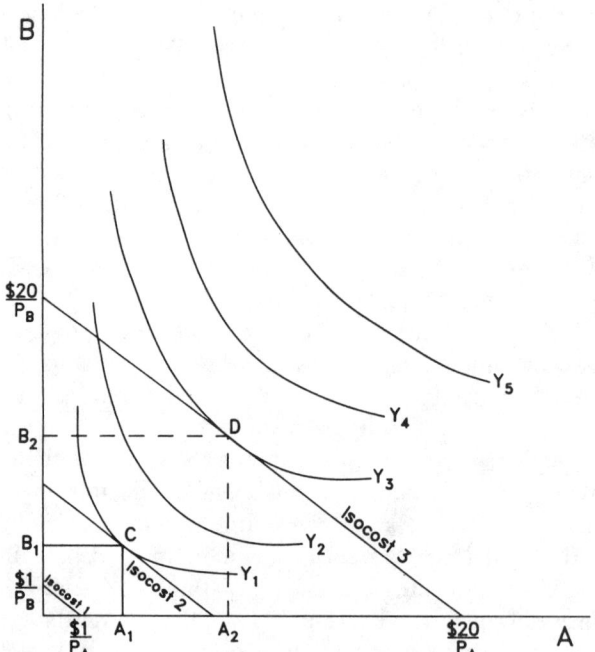

Figure 6.4 Imposing isocost lines on the isoquant map to calculate minimum cost and maximum production optima.

OPTIMUM COMBINATION OF INPUTS

At point C, the slope of isoquant Y_1 ($\text{MRTS}_{AB} = \text{MPP}_A/\text{MPP}_B$) is equal to the slope of the isocost lines [$(\$1/P_B)/(\$1/P_A) = P_A/P_B$] or $\text{MPP}_A/\text{MPP}_B = P_A/P_B$. In conventional production economics, this optimum condition[4] is usually written as $\text{MPP}_A/P_A = \text{MPP}_B/P_B$.

The Maximum Product Optimum The answer to the second question of the optimum input mix to maximize production begins with the isocost lines of Figure 6.4. Suppose our goal is to maximize production from an operating budget of $20. The optimum combination of A and B that we are seeking will correspond to the highest possible output level (highest isoquant) that can be reached with a budget of $20. We begin by establishing the isocost line dictated by a budget of $20 (isocost 3). As shown, isocost 3 becomes tangent to isoquant Y at point D, indicating that the optimum input combination of A_2 and B_2 produces a maximum output of Y_3.[5]

The Expansion Path and the Optimum Production Level

The third question, the optimum production level to maximize net return, is answered by combining all information thus far presented for the general isoquant–isocost model. As mentioned in the preceding graphical example section, we know that the optimum production level will occur on the isoquant map where the following condition is met: $\text{MPP}_A/P_A = \text{MPP}_B/P_B$. This is generally termed the necessary condition for maximum net return when two variable inputs are used to produce one product. The sufficient condition for maximum net return is, of course, that marginal revenue equal marginal cost. In algebraic form, the necessary and sufficient conditions may be combined as follows:

$$\frac{\text{MPP}_A}{P_A} = \frac{\text{MPP}_B}{P_B} = \frac{1}{\text{MC}} = \frac{1}{P_Y} = \frac{1}{\text{MR}}.$$

By definition, MPP_A/P_A equals the inverse of marginal cost, since $\text{MPP}_A = \Delta Y/\Delta A$ and $P_A = \Delta \text{TC}/\Delta A$ or $\text{MPP}_A/P_A = (\Delta Y/\Delta A)/(\Delta \text{TC}/\Delta A) = \Delta Y/\Delta \text{TC} = 1/\text{MC}$. Also, by definition, $1/P_Y = 1/\text{MR}$ in pure competition, since each unit of product must be sold at the going market price. In terms of Figure 6.5, then, the necessary condition is met everywhere along the expansion path that traces out the various points where isocost lines are tangent to the corresponding isoquants. The sufficient condition for maximum net return is only met at one unique point on the expansion path. Mathematically, this point is easily located.[6] In graphical terms, it is more difficult. Graphically finding the correct output level involves moving out along the expansion path until $\text{MC} = \text{MR}$. Outward movement along the expansion path from Y_1 to Y_2, etc. brings a decrease in $1/\text{MC}$ values (an increase in MC) since $\text{MPP}_A/P_A = \text{MPP}_B/P_B$ is satisfied at each point, P_A and P_B are constant, and due to the law of diminishing returns, increases in A and B decrease their MPP values.

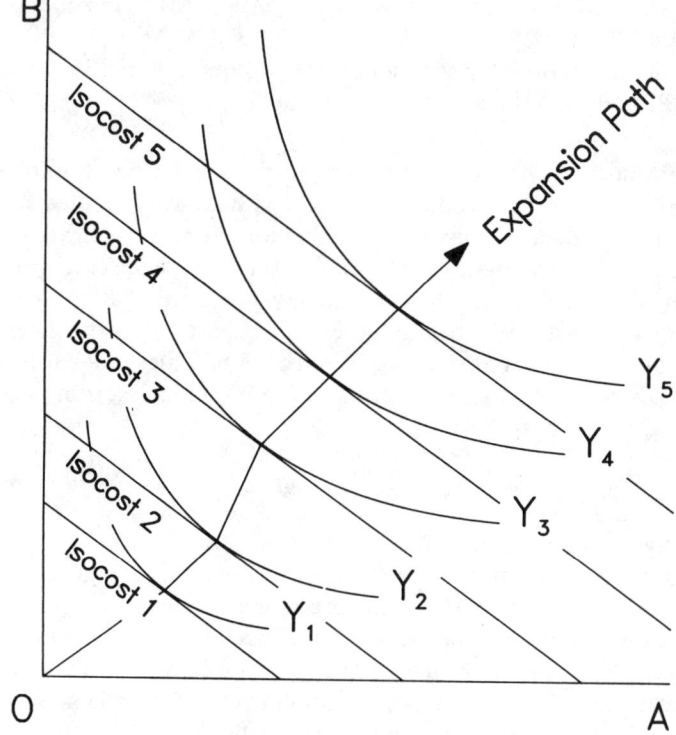

Figure 6.5 Expansion path tracing out the points of tangency between isocost lines and isoquants.

Graphically locating the optimum Y, then, is really a matter of moving along the expansion path until MC has been increased enough to equal MR. Perhaps another graph will help us visualize what is really happening. Figure 6.6 represents a TPP curve for combined inputs of A and B along the expansion path of Figure 6.5, corresponding APP and MPP curves, and corresponding AVC and MC curves derived from the production information. In production Stage II, as we move out along the expansion path of Figure 6.5, we simultaneously move up along the TPP curve, down along the APP and MPP curves, and up along the AVC and MC curves of Figure 6.6. Thus, we can visualize the determination of the optimum output level as a movement out along the expansion path of Figure 6.5 and a corresponding movement up along the MC curve of Figure 6.6 until MC intersects the MR line, identifying Y_0 as the optimum production level.

Input Adjustments to Achieve Optimum Production

The above discussion of optimizing procedures has centered on the identification of the necessary and sufficient conditions for maximum net returns.

OPTIMUM COMBINATION OF INPUTS

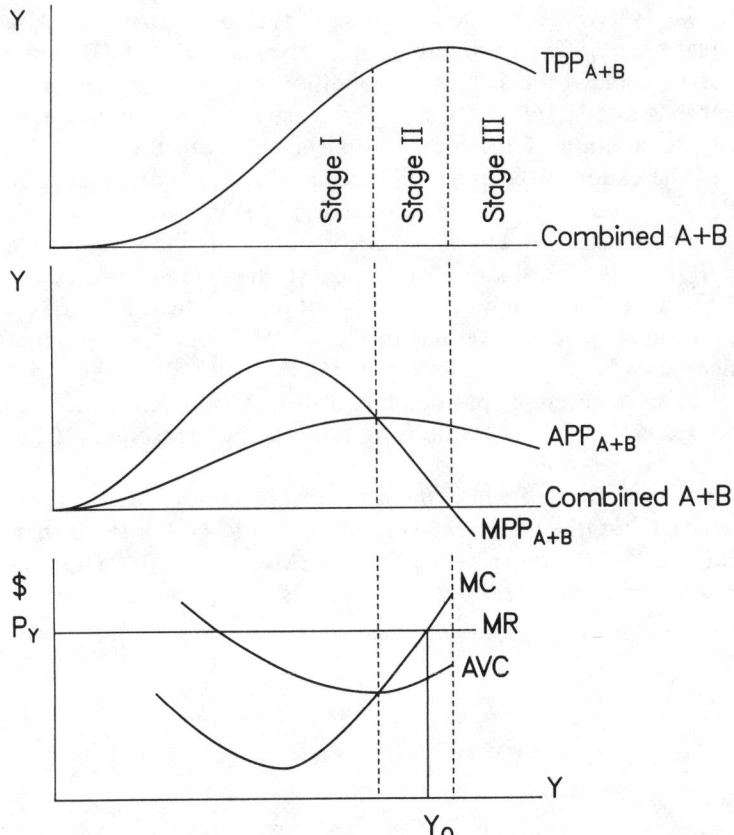

Figure 6.6 Production functions and cost curves representing combined inputs of A and B along the expansion path.

Calculation of optimum input mix and optimum output level has been described in terms of solving algebraic equalities for unknown quantities of variable inputs A and B. But what if the necessary or sufficient conditions do not hold? What adjustments in the use of A or B must be made?

Let us first examine the case where the necessary condition $MPP_A/P_A = MPP_B/P_B$ is not satisfied. Specifically, suppose the following inequality holds: $MPP_A/P_A > MPP_B/P_B$. What adjustments in the use of A and B should be made to reach the optimum input combination to minimize costs (or maximize production)? Inspection of the inequality reveals that the left side of the equation is larger than the right side. Since variable input prices, P_A and P_B, are market dictated and beyond the control of the manager in pure competition, any changes to force the equation to equality must come about by adjusting the MPP values. Since we are operating between the ridge lines on the isoquant map, production is in Stage II with respect to both variable inputs. So what must be done to make the left side of the inequality smaller and the right side larger? To decrease MPP_A, more of variable input A must be used and to increase MPP_B, less of variable input B must be used. Thus the input adjustment to achieve the necessary condition (optimum input mix), involves using more A and less B or "trading" the use of A for the use of B.

A graphical presentation of the interrelationship between the two MPP values may make the rationale behind this mix adjustment more clear. The inequality $MPP_A/P_A > MPP_B/P_B$ may also be written $MPP_A/MPP_B > P_A/P_B$ and from previous definitions we know that MPP_A/MPP_B is the slope of the isoquant while P_A/P_B is the slope of the isocost line. Transferring this information to Figure 6.7, we know that the position on the isoquant (C) described by the inequality lies above and to the left of the isocost–isoquant tangency (D) necessary for the optimum input mix. Visually, then, the input adjustment required to achieve the optimum mix of A and B is also to increase the use of A and decrease the use of B, moving down along the isoquant from C to D.

Let us now examine the case where the necessary condition, $MPP_A/P_A = MPP_B/P_B$ is satisfied but the sufficient condition, $1/MC = 1/MR$ is not. Specifically, suppose $MPP_A/P_A = MPP_B/P_B = 1/MC < 1/MR$. This inequality describes a situation where the mix of inputs being used is optimum

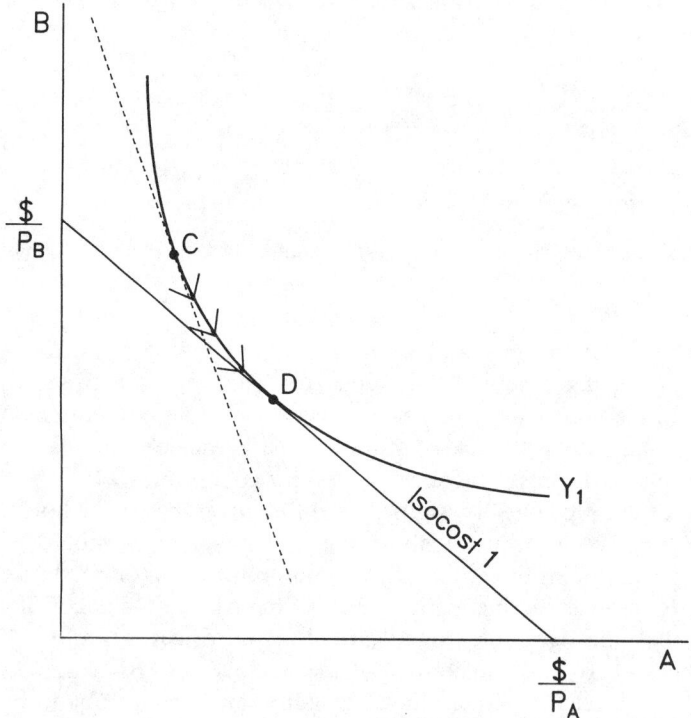

Figure 6.7 Input adjustment to reach the optimum input mix.

but the total quantities of inputs (and the quantity of output) is not. What adjustment in variable input use should be made to satisfy the sufficient condition for maximum net return? Since the inverse of marginal cost, 1/MC is less than the inverse of marginal revenue, 1/MR, MC > MR. In pure competition, MR is set in the market and beyond the control of the operator so any adjustment to this inequality must come through a change in MC. As previously stated, we are operating in the rational stage of production (Stage II) with respect to both variable inputs. Thus, since MC exceeds MR, we know that too much output is being produced and that both inputs, A and B, must be decreased in order to reduce output and decrease MC to where it equals MR. In other words, this inequality describes a situation where production is occurring at a point too far out on the expansion path and both inputs must be reduced (refer to Figure 6.6).

The Numerous Variable Input Case

Thus far this section has discussed the application of the general isoquant–isocost model in both graphical and mathematical terms. Application has been limited to the case of two variable inputs being applied in combination with one set of fixed inputs to produce one product. But what about the situation where more than two variable inputs are combined to produce a product? How would the optimum combination of inputs be determined? Suppose weaner calves are being produced from a cow–calf operation that employs the following variable inputs:

- A = purchased hay,
- B = leased winter range,
- C = leased spring range,
- D = leased aftermath forage,
- E = purchased protein supplement,
- F = hired labor,

in combination with two fixed inputs:

- X = irrigated meadow,
- Z = owner management.

As before, optimum input mix is reached by solving an algebraic equation. The only difference is that in this case the equation is considerably longer than those discussed above. The appropriate necessary condition is established as

$$\frac{\text{MPP}_A}{P_A} = \frac{\text{MPP}_B}{P_B} = \frac{\text{MPP}_C}{P_C} = \frac{\text{MPP}_D}{P_D} = \frac{\text{MPP}_E}{P_E} = \frac{\text{MPP}_F}{P_F} = \frac{1}{\text{MC}};$$

the sufficient condition is again

$$\frac{1}{\text{MC}} = \frac{1}{\text{MR}}.$$

From these two equations, six equations are derived and solved simultaneously for the optimum levels of the six variable inputs.[7]

LITERATURE CITED

Allen, C. L. 1962. *Elementary Mathematics of Price Theory*. Wadsworth, Belmont, CA. 155 pp.

Brokken, R. F. and A. C. Bywater. 1982. Application of isoquant analysis to forage: grain ratios in cattle feeding. *J. An. Sci.* 54:463–472.

Brokken, R. F., T. H. Hammonds, D. A. Dinius, and J. Valpey. 1976. Framework for economic analysis of grain versus harvested roughage for feedlot cattle. *Amer. J. Agr. Econ.* 58:245.

Burt, O. R. 1978. On the statistical estimation of isoquants and their role in livestock production decisions. *Amer. J. Agr. Econ.* 60:519.

Doll, J. P. and F. Orazem. 1978. *Production Economics—Theory with Applications*. Wiley, New York. 406 pp.

Heady, E. O. 1952. *Economics of Agricultural Production and Resource Use*. Prentice-Hall, Englewood Cliffs, NJ. 850 pp.

Heady, E. O. and J. I. Dillon, 1961. *Agricultural Production Functions*. Iowa State Univ. Press, Ames. 642 pp.

Heady, E. O. and John Pesek. 1954. A fertilizer production surface. *J. Farm. Econ.* 36:466–482.

Heady, E. O., G. P. Roehrkasse, W. Woods, and J. M. Scholl. 1963. Beef-cattle production functions in forage utilization. Iowa Agricultural and Home Economics Experiment Station Research Bulletin No. 517.

TEXT NOTES

1. For the sake of simplicity, it will be assumed that purchase and sale prices of steers are both $0.91 per pound. Thus there is no price spread between light and heavy weight animals.
2. In mathematical notation, $\text{MRTS}_{AB} = dB/dA$, where $B = f(A|Y)$.
3. $\text{MPP}_A/\text{MPP}_B = (\Delta Y/\Delta A)/(\Delta Y/\Delta B) = \Delta B/\Delta A = \text{MRTS}_{AB}$.
4. In mathematical terms, optimization is achieved by solving the following equations (1) $\text{MPP}_A/P_A = \text{MPP}_B/P_B$ and (2) $Y = f(A, B)$ simultaneously for A and B, the minimum cost combination of variable inputs.
5. In mathematical terms, optimization is achieved by solving the following equations (1) $\text{MPP}_A/P_A = \text{MPP}_B/P_B$ and (2) $P_A \cdot A + P_B \cdot B = \$20$ simultaneously for A and B, the maximum output combination.
6. In mathematical terms, optimization involves solving the following equation for the optimum levels of A and B, which also sets the maximum net return level of Y: $\text{MPP}_A/P_A = \text{MPP}_B/P_B = 1/P_Y$. Alternatively, rewriting to produce two equations to correspond to the two unknowns, A and B: $(\text{MPP}_A \cdot P_Y)/P_A = 1$ and $(\text{MPP}_B \cdot$

$P_Y)/P_B = 1$ or $\text{VMP}_A/P_A = 1$ and $\text{VMP}_B/P_B = 1$.

7 In mathematical terms, the production function may be written $Y = f(A, B, C, D, E, F | X, Z)$ and the following six equations are solved simultaneously to yield the optimum quantities of the six inputs:

$$P_Y \cdot \frac{\partial Y}{\partial A} = 1,$$
$$P_Y \cdot \frac{\partial Y}{\partial B} = 1,$$
$$\vdots$$
$$P_Y \cdot \frac{\partial Y}{\partial F} = 1.$$

Chapter 7

Optimum Combination of Outputs—The Economics of Multiple Use

In range management, as well as natural resource management in general, some of the more common and most difficult decision problems involve allocation of fixed[1] land resources among competing uses. How much of a particular tract of rangeland should be devoted to livestock grazing and how much should be set aside for big game habitat? What proportion of a high country area should be open for recreational activities and what proportion should remain closed to protect important watersheds? Questions like these may not have definite answers nor even acceptable means of making justifiable approximations. However, in any land use decision for which the alternative products have established market prices (as do livestock forage, timber, culinary and irrigation water as well as some recreational pursuits), production economics principles can be used to determine the optimum allocation to maximize combined net return to the fixed land resource. Even for those natural resource products that we all agree are crucial (such as watershed protection and preservation of wildlife habitat) but which have no established prices, principles of production economics can be used to calculate what the minimum values of such products would have to be in order for a particular land use allocation to be rational.

OPTIMUM COMBINATION OF OUTPUTS

Interactions between alternative land uses are usually considered to be of four general types: (1) competitive, (2) complementary, (3) supplementary, and (4) antagonistic (Heady, 1952). Each of these types of trade-off relationships will be discussed in turn.

COMPETITIVE PRODUCTS

As the name implies, competitive products are those for which an increase in output of one product brings a corresponding decrease in output of a second product. Competition between alternative products may occur at either constant or increasing rates of substitution.

Competition at Constant Rates of Substitution

Suppose we have 100 acres of native wet meadow that can be used to produce hay or, alternatively, can be harvested directly by livestock as grazed forage. If devoted to hay production, each acre of meadow will yield 1 ton of hay annually, while if grazed, annual sustained yield production amounts to 2 AUM per acre (Figure 7.1). Thus, as shown in the "production functions"[2] for hay and forage in the upper portion of Figure 7.1, if the entire 100 acres is used for hay production, 100 tons of hay are produced. Alternatively, if the entire meadow is devoted to grazing, 200 AUM are produced. These two quantities of production, 100 tons and 200 AUM, form the endpoints on the transformation curve (also called trade-off or production possibility curve) shown in the lower portion of Figure 7.1. The term *transformation* describes how hay production may be transformed into forage production by changing the land use allocation of our meadow. For example, beginning at the point corresponding to 100 acres and 100 tons in the upper left graph in Figure 7.1, if the acreage devoted to hay is reduced to 50 acres, hay production drops to 50 tons. But simultaneously, as shown in the upper right graph of Figure 7.1, the 50 acres released from hay production are now available for forage production allowing 100 AUM to be produced. Thus, 50 tons of hay production have been transformed or traded for 100 AUM of forage production as shown in the lower graph of Figure 7.1. This transformation curve shows all possible combinations of hay and forage that can be produced on our 100-acre meadow. Since both production functions giving rise to the transformation curve are linear, the transformation curve itself is also linear, and the two products are said to compete or substitute for each other at a constant rate. The rate of substitution may be defined as the slope of the transformation curve, which, in our example, remains constant at 2 AUM/1 ton or 2.[3]

The transformation curve of Figure 7.1 shows what is biologically possible on our 100 acres of wet meadow. But how should the land be allocated between the two uses in order to maximize net return to our 100 acres? As with all optimization decisions, the choice of how many acres to devote to hay production and how many to graze must be based on product price in

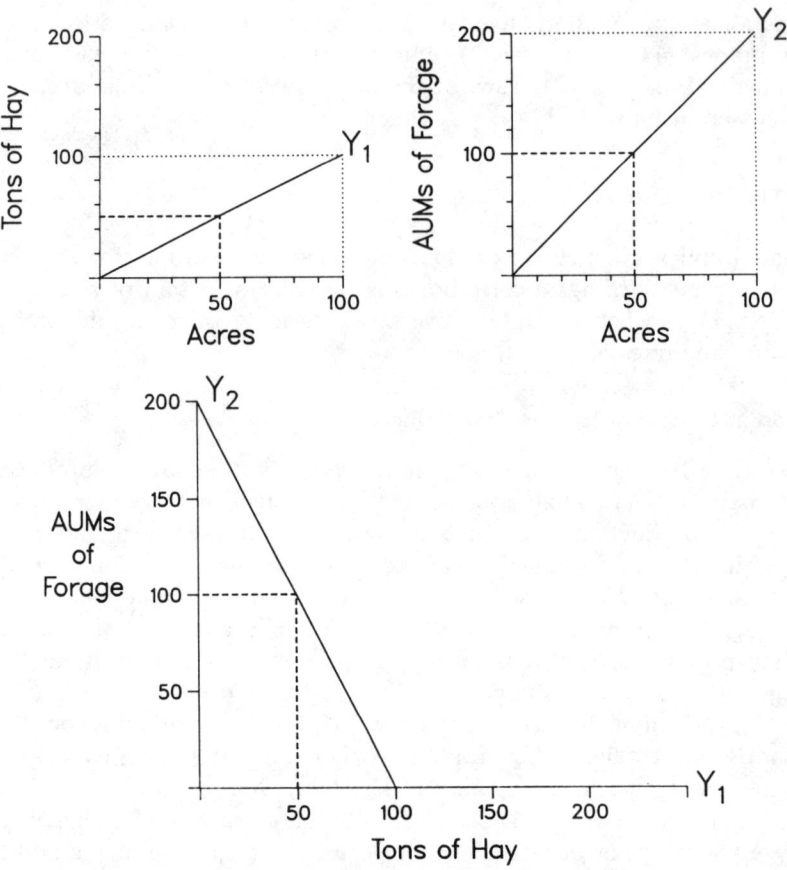

Figure 7.1 Production functions and resulting transformation curve for production of hay or forage from 100 acres of wet meadow.

combination with the production function. In this case the optimum mix of products is determined by combining net product prices with the transformation curve. Suppose hay has a market value of $40 per ton and harvesting and stacking costs total $15 per ton, giving a net hay price of $25 per ton. Suppose further that meadow forage may be leased at $8 per AUM and the costs of herding, supervision, water, salting, and improvement maintenance amount to $3 per AUM resulting in a net forage price of $5 per AUM. Based on these net prices, if the entire meadow is grazed by livestock, it will yield a net return of only $1000 whereas if the entire 100 acres are used to produce hay net return will total $2500. Clearly exclusive hay production is far superior to exclusive grazing. Since the transformation curve is linear, exclusive hay production also yields a higher net return than any possible mixture of hay and grazing. For example, if one half of the meadow is devoted to hay and the other half grazed, net return is $1250 from hay and $500 from forage for a total net return of only $1750.

OPTIMUM COMBINATION OF OUTPUTS

An easier and faster method of determining the optimum product combination is to again apply the now familiar principle of marginality by imposing an isorevenue line on the transformation curve of Figure 7.1. In Figure 7.2 the transformation curve appears as line AB and again shows all combinations of hay and forage that our 100-acre meadow is capable of producing. But our goal is to maximize net return so we need another line (line CD) to represent this parameter. Line CD is positioned in the graph of Figure 7.2 by answering two questions. First, how many tons of hay must be sold to yield $500[4] net revenue? The answer, $500/($25 per ton) = 20 tons, is plotted as endpoint D of line CD. Second, how many AUM of forage must be sold to yield $500 net revenue? Point C, calculated as $500/($5 per AUM), is plotted as the other endpoint of line CD. A straight line is now drawn between points C and D, resulting in inverse[5] price line of isorevenue line CD that shows all possible combinations of hay and forage capable of yielding $500 net return. However, our goal is not to produce $500 net return. It is to maximize net return consistent with the biological capability of our 100 acres. Thus in terms of Figure 7.2, we want to move line CD out as far as possible (keeping it always parallel to the original line CD in order to hold the inverse price ratio constant at $P_{Y_1}/P_{Y_2} = \$25/\$5$) while just maintaining contact with line AB. Line EB represents the highest net revenue that our 100-acre meadow is capable of producing. Point B, where line EB is tangent to the line AB, marks the optimum combination of hay and forage production. Due to the linear nature of the transformation curve our land allocation decision is an "all or nothing"

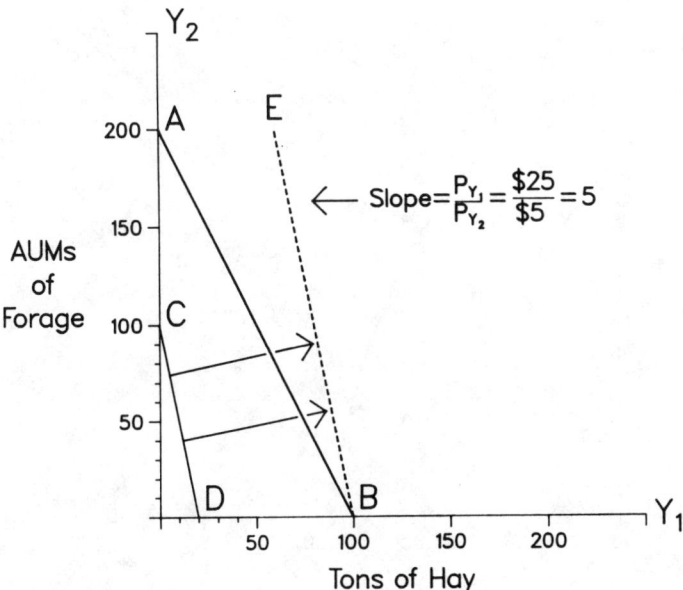

Figure 7.2 Determination of optimum combination of forage and hay by imposing an inverse price line on a transformation curve.

or "corner solution" at point B and maximum net return again totals \$2500 (\$25 × 100 tons). Any inverse price ratio, P_{Y_1}/P_{Y_2} that exceeds the slope of line AB (20/10 = 2) would also give point B as the optimum land use, while any inverse price ratio less than 2 would dictate point A (exclusive forage production). In the unlikely, but possible, situation where the inverse price ratio equalled the slope of the linear transformation curve (2), all possible combinations of hay and forage production would yield identical amounts of net revenue.

The linear transformation function represents the most simple product trade-off possible and probably applies only in cases where cropland is being allocated between the production of two competing but *noninteracting* crops. We will next examine the more common (and more complex) relationships between land uses that do interact.

Competition at Increasing Rates of Substitution

The second type of competitive trade-off between alternative products is one in which increased output of one product must be matched by successively larger decreases in output of a second product. Suppose we are managing a 1000-acre tract of rangeland and our sustained yield production possibilities for calf and lamb are portrayed by curve $ABCD$ in Figure 7.3. At point A our 1000 acres of range are allocated exclusively to sheep grazing and 7500 pounds of lamb are produced. Exclusive cattle grazing occurs at point D resulting in 6000 pounds of calf production. Between points A and D the range is grazed by a

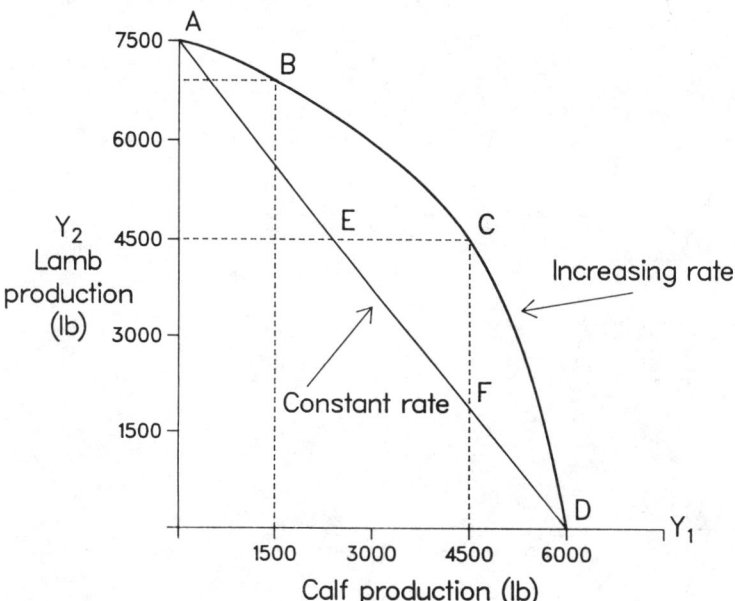

Figure 7.3 Transformation curves for two products that compete at constant and increasing rates of substitution.

mixture of the two, yielding various combinations of both calf and lamb production.

If our operation were currently devoted exclusively to sheep production (point A), a few cattle could be substituted for a small number of sheep (point B), allowing a 1500-pound increase in calf production, while reducing lamb production by only 700 pounds (from 7500 to 6800 pounds). However, if we were currently operating at point C, a 1500-pound increase in calf production (a movement to point D) would reduce lamb production by 4500 pounds. Thus the rate of substitution of calf production for sheep production increases by more than six times between points A and C.

The same relationship is evident if we begin at point D with exclusive cattle production and substitute sheep for cattle in a movement along the transformation curve back toward point A. The first 4500-pound increase in lamb production requires a sacrifice of only 1500 pounds of calf production (a movement from point D to point C), while the next 3000-pound lamb production increase requires a 4500-pound decrease in calf production (a movement from point C to point A).

As mentioned, the rate of substitution between the two products is more formally referred to as the *marginal rate of substitution* (MRS) of Y_1 for Y_2[6] and is defined as

$$\text{MRS}_{Y_1 Y_2} = \frac{Y_2}{Y_1}.$$

$\text{MRS}_{Y_1 Y_2}$ represents the slope of the transformation curve at a particular point on the curve and may be thought of as (quantity of Y_2 that must be given up)/(quantity of Y_1 gained). As will be seen, it is often helpful to define $\text{MRS}_{Y_1 Y_2}$ in terms of marginal physical products for each of the two outputs. Referring to Figure 7.1, MPP for fixed input X (land) used to produce Y_1 may be written

$$\text{MPP}_{X-Y_1} = \frac{\Delta Y_1}{\Delta X}.$$

Similarly, $\text{MPP}_{X-Y_2} = \Delta Y_2 / \Delta X$. Thus $\text{MRS}_{Y_1 Y_2}$ may be defined as

$$\text{MRS}_{Y_1 Y_2} = \frac{\text{MPP}_{X-Y_2}}{\text{MPP}_{X-Y_1}} = \frac{\Delta Y_2 / \Delta X}{\Delta Y_1 / \Delta X} = \frac{\Delta Y_2}{\Delta Y_1},$$

which is identical to our original definition of $\text{MRS}_{Y_1 Y_2}$.

Biological Interpretation of Competitive Products Why might the trade-off relationship between cattle and sheep be shaped like the transformation curve $ABCD$ in Figure 7.3 rather than like straight line $AEFD$? For what reason might we expect the substitution of cattle for sheep to occur at an increasing rather than a constant rate? In an attempt to answer these questions let us again suppose we have an exclusive sheep operation producing at point A in Figure 7.3. At point A the $\text{MRS}_{Y_1 Y_2}$ as measured by the slope of transformation curve $ABCD$ is relatively flat, which means that the ratio of amount of lamb production given up to amount of calf production gained is relatively small. This seems consistent with what we would expect under an actual range situation. On a range that has previously been grazed only by sheep it is likely that a few cattle could be added with little detrimental impact on sheep production simply because sheep and cattle do not prefer the same forage plant species. During the spring grazing season, for example, sheep might select primarily forbs while cattle might concentrate on grasses. Thus the first few head of cattle added to the sheep operation graze almost entirely on plants that the sheep were not using. However, as the number of cattle grazing our 1000-acre range increases in proportion to the number of sheep occupying the area, the cost of increased calf production in terms of lamb production lost might become higher and higher (as shown by the ever increasing slope of curve $ABCD$ as we proceed from point A down the curve to points B and C). Such an increase in substitution rate would be expected since the higher proportion of cattle to sheep might well force the cattle to compete directly with the sheep for the same plants. In the extreme case, as we move from point C toward point D, more and more sheep are removed from the range but competition among the ever increasing number of cattle for a limited quantity of preferred plants causes increased calf production to taper off and to stabilize at 6000 pounds (point D) as the last sheep are removed from the range and lamb production ceases altogether. The same phenomenon would be expected if we began with an exclusive cattle operation and substituted sheep for cattle in a movement back along curve $ABCD$.

Biologically speaking, then, what is the potential advantage of common use of the range by both sheep and cattle over either kind of animal grazing alone? If the trade-off curve between sheep and cattle were represented by the linear curve $AEFD$, we would expect our range to be capable of producing 4500 pounds of lamb in combination with about 2400 pounds of calf (point E, Figure 7.3). Instead, due to competition between the two animals at an increasing rate of substitution, we find that we can produce 4500 pounds of calf (point C), an unexpected "bonus" of 2100 pounds. Likewise, if the trade-off curve were linear, we would expect that we could product 4500 pounds of calf in combination with about 1875 pounds of lamb (point F). Instead, due to competition at an increasing substitution rate we can produce 4500 pounds of lamb, a bonus of 2625 pounds.

From a purely biological standpoint, then, the advantage of common use over single use grazing is that more meat can be produced from a given rangeland area because of more complete plant utilization while still not exceeding sustained yield carrying capacity (Cook, 1954; Smith, 1965). We next turn to an economic optimization procedure based on transformation curve $ABCD$ of Figure 7.3.

OPTIMUM COMBINATION OF OUTPUTS

Economic Interpretation of Competitive Products Our optimization procedure again consists of imposing an isorevenue line on the transformation curve AOD of Figure 7.4. Isorevenue line BC shows all possible combinations of calf and lamb production capable of yielding $1200 net return. Points B and C were calculated by dividing $1200 by the net prices of lamb and calf, $0.50 and $0.40 per pound, respectively. As before, our goal is to maximize combined calf and lamb net return from our 1000 acres of rangeland, subject to not exceeding sustained yield livestock production as represented by curve AOD. In graphical terms, our goal is to find the straight line that is parallel to isorevenue line BC and also tangent to transformation curve AOD. Line EOF is the isorevenue line we are seeking and its tangency with the transformation curve at point O marks the optimum product combination (3700-pound calf production and 5400-pound lamb production) for maximum net revenue. At point O the slope of the isorevenue curve $(P_{Y_1}/P_{Y_2} = \$0.40/\$0.50 = 0.8)$. Thus at point O the following equality holds:

$$\text{MRS}_{Y_1 Y_2} = \frac{P_{Y_1}}{P_{Y_2}}.$$

By definition this equality can be rewritten

$$\frac{\text{MPP}_{Y_2}}{\text{MPP}_{Y_1}} = \frac{P_{Y_1}}{P_{Y_2}}$$

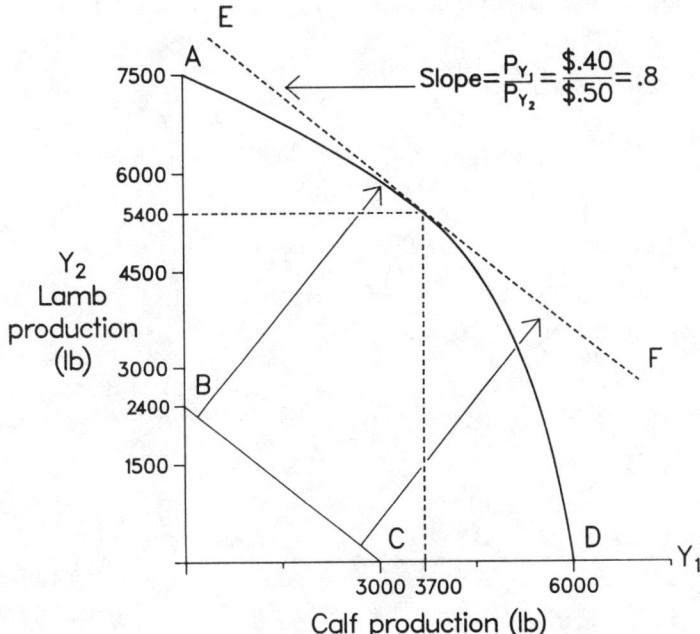

Figure 7.4 Optimum combination of two products that compete at an increasing rate of substitution.

or, rearranging terms,

$$(\text{MPP}_{Y_2})(P_{Y_2}) = (\text{MPP}_{Y_1})(P_{Y_1})$$

or, by definition of VMP,

$$\text{VMP}_{Y_2} = \text{VMP}_{Y_1}.$$

In other words, at point O the value of lamb production given up is exactly equal to the value of calf production gained. At any point lying to the left of point O the value of lamb production given up (VMP_{Y_2}) is less than the value of calf production gained[7] and net return could be increased by allocating more of our 1000 acres to cattle production and less to sheep production. At any point on curve AOD to the right of point O, net return could be increased by expanding lamb production and reducing calf production. Only at point O is there nothing to be gained by changing the allocation of our range between the two types of livestock, and we have found the optimum combination of sheep and cattle production.

COMPLEMENTARY PRODUCTS

A complementary trade-off relationship may be defined as one in which successive increases in output of one product result in simultaneous increases in output of a second product (Workman, 1975). A hypothetical complemen-

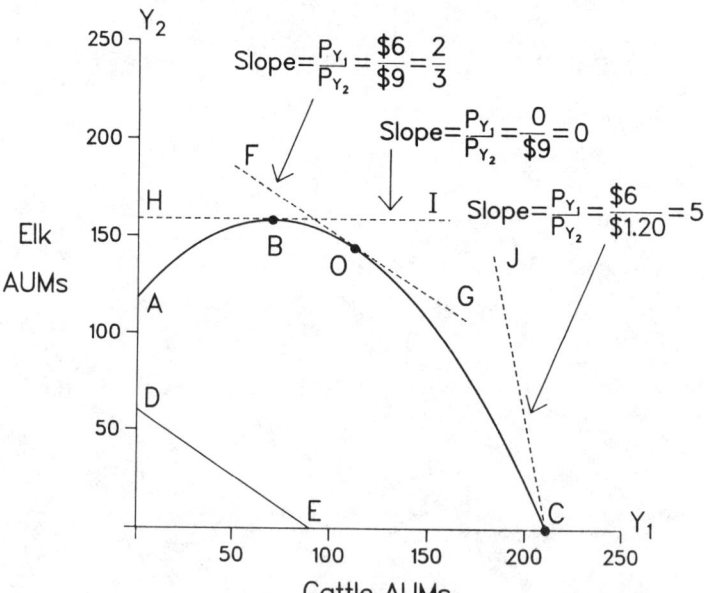

Figure 7.5 Optimum combinations of two complementary products at three different price ratios.

tary transformation curve for cattle and elk forage produced on a particular tract of rangeland is shown in Figure 7.5.

Biological Interpretation of Complementary Products

The AB region of transformation curve $ABOC$ demonstrates a complementary relationship between forage production for early summer cattle use and winter elk use (Smith and Doell, 1968). At point A, only elk utilize the range. Moving along the curve from A to B, increasing numbers of cattle begin to graze the range in the summer in combination with the elk already present during the winter. But unlike the competitive relationship portrayed in Figure 7.4, increased production of summer cattle AUM are matched by simultaneous increases in winter forage available for elk. Between A and B, increasing numbers of cattle utilize more of the grasses that they prefer over shrubs. This allows the shrubs to better compete for space, light, moisture, and nutrients resulting in greater carrying capacity for elk.

Between points B and C the trade-off relationship is competitive, with increases in production of cattle AUM matched by simultaneous decreases in elk forage. Throughout the BC portion of the curve, increasing numbers of cattle are forced to compete more and more directly with the elk for forage. As the grasses preferred by the cattle become scarce, the shrubs preferred as winter forage by the elk furnish an increasing portion of the cattle summer diet. In the extreme case at point C, all forage has been utilized by summer cattle grazing and none remains for winter elk use.

Economic Interpretation of Complementary Products

Suppose that net market prices are $6 per AUM for cattle forage and $9 per AUM for elk forage. The optimization procedure begins with the positioning of the isorevenue line DE. In order to produce $540 net return (or some other arbitrary amount easily divisible by cattle and elk forage prices), how many AUM of cattle forage must be produced? The answer ($540/$6 = 90 AUM) is located as point E. Similarly, point D illustrates the quantity of elk AUM required to produce $540 ($540/$9 = 60 AUM). Isorevenue line FG, parallel to isorevenue line DE, is now drawn tangent to transformation curve $ABOC$, identifying point O (approximately 125 cattle AUM and 140 elk AUM) as the optimum output combination for maximum net revenue.

Calculating Nonmarket Prices Justifying Allocation Decisions

As previously acknowledged, numerous important rangeland products such as wildlife and recreation benefits lack established market prices and are extremely difficult to compare with livestock forage, water, and other products. In the example of Figure 7.5, suppose the market price of cattle forage is $6 per AUM but no comparable price exists for elk AUM. Further suppose that the range manager has decided to allocate all available forage to summer cattle grazing, leaving none for winter elk use (point C). What price is implied for elk forage by this decision? As shown in Figure 7.5, the slope of the transformation curve at point C (where the isorevenue line must come tangent in order to dictate point C as the optimum production combination) is about $5:1$.[8] Thus the slope of isorevenue line JC must also equal $5:1$ and since the price of cattle forage (P_{Y_1}) is known to be $6 per AUM, the implied price of elk forage (P_{Y_2}) must equal $1.20 per AUM ($P_{Y_1}/P_{Y_2} = \$6/\$1.20 = 5/1$). If the allocation of all forage available to cattle use is an economically rational decision, the implication is that an elk AUM is worth only $\frac{1}{5}$ as much as an AUM used by cattle.

Finally, suppose our range manager is accused of being biased in favor of livestock by a sportmen's representative, who maintains that the range should be allocated to exclusive elk use during the winter (point A). In the sportsman's opinion, forage produced for livestock use has no value, while forage to be used by wintering elk has a value at least as high as the $9 per AUM price previously mentioned. But what is the economic optimum allocation of forage between cattle and elk at these prices? Isorevenue line HI is based on prices of $0 per cattle AUM and $9 (or any positive dollar amount) per elk AUM. Line HI is tangent to the transformation curve at point B, indicating an optimum output mix of 70 AUM of cattle forage and about 156 AUM of elk forage. Even if cattle AUM have no value, as long as elk AUM do have some positive value, it pays to allocate 70 AUM to cattle grazing because of the beneficial effects of cattle grazing on forage production for elk. Thus, in the example of Figure 7.5, as long as either elk or cattle AUM have a price greater than zero, optimum allocation cannot occur at point A (exclusive elk use). Nor would single use by cattle be an optimal land use allocation unless the price of cattle forage exceeded elk forage price by a ratio of $5:1$.

Both of the above examples argue convincingly for multiple use (some cattle and some elk) rather than exclusive use by either animal. Even in the absence of market prices for some rangeland products, transformation curves can be used to calculate what the elusive product values would have to be in order for certain land allocation decisions to be economically sound. For example, in the absence of a price for elk AUM, it could be calculated that in order for point O to be an optimum land use mix, elk AUM would have to be priced at $9 each. The range manager could next appeal to his various publics for input as to whether the $9 figure were too high (suggesting that more forage should be allocated to cattle and less to elk) or too low (suggesting that the range should be used more by elk and less by cattle). It should be recognized, however, that derivation of trade-off curves between various rangeland products is immensely more difficult than application and interpretation of such curves after they have already been derived (Cook, 1954; Hopkin, 1954; Smith, 1965; Lewis, 1970).

SUPPLEMENTARY PRODUCTS

The third general type of trade-off is the supplementary relationship that may be defined as the production situation in which successive increases in output of one product have no effect on output of a second product. A hypothetical

OPTIMUM COMBINATION OF OUTPUTS

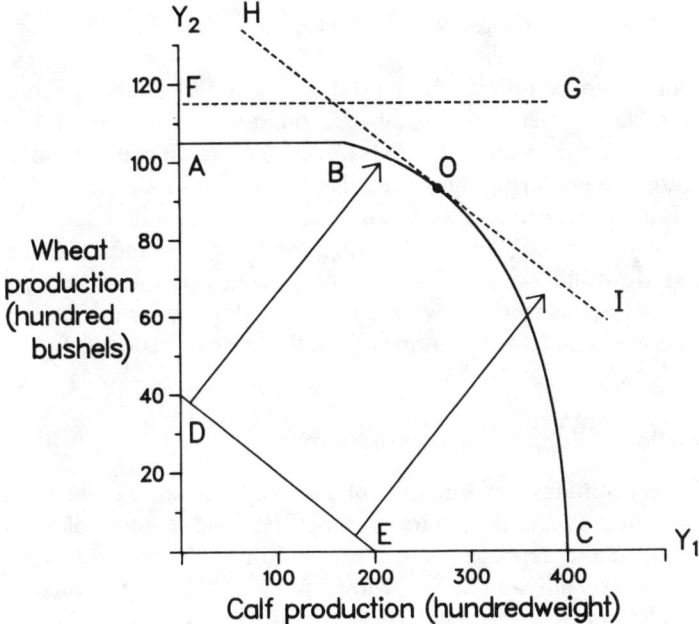

Figure 7.6 Optimum combinations of two supplementary products at two different price ratios.

supplementary relationship between cattle and wheat production is shown in Figure 7.6.

Biological Interpretation of Supplementary Products

The AB region of transformation curve $ABOC$ illustrates "supplementarity" between cattle and wheat production on a particular ranching unit. At point A the entire operation (land, labor, and capital) is devoted to exclusive wheat production. Moving from point A to point B, increasing amounts of calf production are added to the wheat enterprise. However, up to point B (160-cwt calf production) calf production is accomplished by utilizing resources (land, labor, and operating capital) that are not used by the wheat enterprise. It is only after the quantity of calf production exceeds 160 cwt at point B that competition between the two products occurs. To the right of point B increased calf production is possible only at the cost of reduced wheat production. Throughout the competitive region of the curve (BOC), increased calf production requires resources also needed by the wheat enterprise. Perhaps labor required at calving time subtracts directly from that available for spring planting and cultivating. Operating capital used to purchase protein meal and hay may directly detract from funds required for seed, fertilizer, and fuel. As the cattle operation expands, a range seeding may be required on land normally planted with wheat. To the left of point B, however, labor, forage, and operating funds required for cattle production draw only on surplus resources that would not be used for wheat production even if the livestock operation ceased.

Numerous examples of supplementary products occur throughout range and natural resource management. For instance, moderate levels of upland or big game hunting might occur in the fall without detracting from production of cattle grazing the same range during the spring. Decreased livestock productivity in the form of four-wheel drive damage and destruction of water facilities and fences normally occurs only when hunting pressure becomes intense. Similarly, a modest downhill ski development can often operate with only natural parks and meadows as runs. However, as the ski development becomes larger, runs must be expanded by tree removal at the cost of reduced future timber production.

Economic Interpretation of Supplementary Products

Determination of the optimum combination of two supplementary products proceeds exactly as already described for competitive and complementary products. Isorevenue line *DE* represents a net revenue of $100 assuming a net calf price of $0.50 per pound and a net wheat price of $2.50 per bushel. Parallel to the line *DE*, isorevenue line *HI* becomes tangent to the transformation curve at point *O*, indicating an optimum product combination of just over 240 cwt of calf production and 9000 bushels of wheat.

It is worth noting that when supplementary substitution occurs between two products, the optimum production combination will almost always occur in the competitive region of the transformation function. Close inspection of Figure 7.6 reveals that at almost any possible combination of wheat and calf prices, the resulting price line becomes tangent somewhere between points *B* and *C*.[9] Even if the calf to wheat price ratio is very low, tangency still occurs at point *B*, indicating that rather than an exclusive wheat enterprise, 160 cwt of calves should be produced in combination with 10,500 bushels of wheat. Even in the extreme case of a calf price of zero while wheat price remained at some positive value, isorevenue line *FG* would become tangent to the transformation curve at all points on the supplementary region (*AB*). Thus, even if calf production had no net market value, it would still be a matter of indifference as to whether any calves should be produced (calf production would neither add to nor subtract from net return).

ANTAGONISTIC PRODUCTS

An antagonistic relationship between two products is said to occur when successive increases in one product bring correspondingly smaller and smaller decreases in the second product. Thus, as with the competitive regions of the various types of trade-off relationships already discussed, $MRS_{Y_1Y_2}$ is negative.[10] But unlike the preceding competitive substitutions, in which $MRS_{Y_1Y_2}$ increased from left to right along the transformation curve, an antagonistic

OPTIMUM COMBINATION OF OUTPUTS

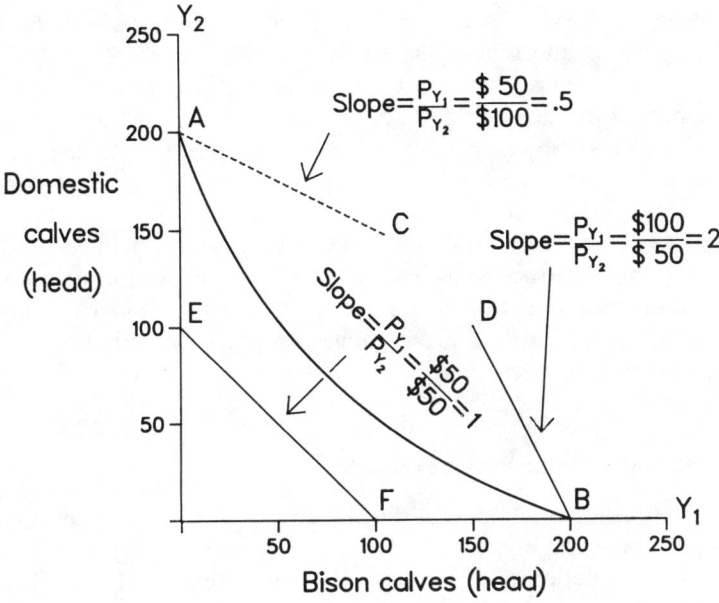

Figure 7.7 Optimum combinations of two antagonistic products at three different price ratios.

trade-off is characterized by a decreasing $MRS_{Y_1Y_2}$. A hypothetical antagonistic relationship between domestic cattle and native American bison grazing the same rangeland parcel is illustrated in Figure 7.7.

Biological Interpretation of Antagonistic Products

Antagonistic transformation curve AB indicates that bison and cattle compete when grazing the same rangeland area but that the competition takes place at a decreasing substitution rate. As shown by the decreasing slope ($MRS_{Y_1Y_2}$) of curve AB, the quantity of domestic calf production that must be sacrificed in order to produce one additional unit of bison calf production decreases along the curve between A and B. What biological mechanism could account for this unusual interaction between two rangeland products? A disease to which both animals are susceptible is the most likely cause of antagonistic product substitution. At point A the range is grazed exclusively by cattle that are brucellosis-free. Moving down and to the right along the curve from point A a few head of bison begin to graze the range in direct competition with the cattle already present. But the competition between the two kinds of animals involves more than just forage utilization. The very first bison to appear on the scene expose the cattle herd to brucellosis, sharply reducing domestic calf production. Once a source of this contagious disease is present, additional bison compete with the cattle only in terms of forage and the cost of successive bison additions (in terms of reduced cattle production) becomes smaller and smaller.

The same mechanism explains the convex shape of the transformation curve beginning at point B and moving up and to the left toward point A. If point B represents exclusive use of the range by a disease-free herd of bison, addition of a few brucellosis-carrying domestic cattle would drastically reduce production of bison calves. However, once initial exposure to the disease has occurred, additional cattle compete with bison production only in terms of forage utilization.

Other examples of antagonistic trade-offs in natural resource management include domestic and bighorn sheep (both carriers of liver flukes) and perhaps even the adverse interaction between snowmobile enthusiasts and cross-country skiers.[11] The classical agricultural production economics example involves mutually contagious diseases of chickens and turkeys when the two birds are raised in close proximity (Heady, 1952).

Economic Interpretation of Antagonistic Products

Suppose net prices of bison and domestic calves are $100 and $50, respectively (isorevenue line DB). Line DB becomes tangent to the transformation curve at point B, giving a corner solution and indicating that the optimum allocation of the range is exclusive bison production.

Next, suppose net prices for the two animals are reversed: $100 for domestic calves and $50 for bison calves (isorevenue line AC). The optimum is again a corner solution, this time at point A, exclusive domestic calf production.

Finally suppose net prices are equal at $50 per calf (isorevenue line EF). If line EF is moved out as far as possible while still remaining in contact with curve AB, there are two points of tangency, A and B. Thus if product prices are equal, it is a matter of indifference which of the two animals will be produced, but clearly the choice will be one of single rather than common use. Of course if the bison price were even one cent higher than the price of domestic calves, exclusive bison production would be the clear choice and vice versa. It should be noted that of the four types of trade-off curves discussed, only antagonistic relationships always result in single use. In contrast, multiple use is the usual solution when the choice is between competitive, complementary, or supplementary products.

LITERATURE CITED

Cook, C. W. 1954. Common use of summer range by sheep and cattle. *J. Range Manage.* 7:10–13.

Heady, E. O. 1952. *Economics of Agricultural Production and Resource Use.* Prentice-Hall, Englewood Cliffs, NJ. 850 pp.

Hopkin, J. A. 1954. Economic criteria for determining optimum use of summer range by sheep and cattle. *J. Range Manage.* 7:170–175.

Lewis, E. S. 1970. Design of a study to reveal a workable cattle–sheep substitution curve. Plan B. M.S. report (Unpublished), Range Economics, Utah State University, Logan. 47 pp.

Smith, A. D. 1965. Determining common use grazing capacities by application of the key species concept. *J. Range Manage.* 18:196–201.

Smith, A. D. and D. D. Doell. 1968. *Guide to Allocating Forage Between Cattle and Big Game on Big Game Winter Range.* Utah State Division of Fish and Game Publication No. 68-11. 32 pp.

Workman, John P. 1975. Wildlife and recreation on U.S. rangelands—the economic aspects. In *Arid Shrublands, Proceedings of the Third Workshop of the United States/Australia Rangelands Panel*, Society for Range Management. Tucson, Arizona, April *4, 1973. Pp.* 131–134.

TEXT NOTES

1. Optimum allocation of limited but variable resources among alternative uses is most easily accomplished by the VMP method discussed in Chapter 5.
2. The term "production function" is usually used to describe the relationship between quantity produced and application of various amounts of variable inputs. Here we are using the term to describe the relationship between quantity produced and allocation of various amounts of fixed input (land).
3. More formally, the slope of the transformation curve is defined as the marginal rate of substitution of Y_1 for Y_2 ($MRS_{Y_1 Y_2}$) = $\Delta Y_2 / \Delta Y_1$ = $(200 - 0 \text{ AUM})/(0 - 100 \text{ tons})$ = -2.
4. $500 is an arbitrary choice based on ease of division by hay and forage prices. Other net return amounts such as $25, $50, or $1000 would work equally well.
5. Line CD is an inverse price line with respect to the transformation curve AB. The slope of the transformation line AB equals (AUM of forage)/(tons of hay) whereas the slope of the price line CD is the inverse of (forage price)/(hay price) or slope of CD = (hay price)/(forage price) = P_{Y_1}/P_{Y_2}.
6. Mathematically, $MRS_{Y_1 Y_2} = dY_2/dY_1$ where $Y_2 = f(Y_1)$ represents the functional form of the transformation curve.
7. As apparent in Figure 7.4, at any point on the transformation curve to the left of O, the slope of the transformation curve ($MRS_{Y_1 Y_2} = MPP_{Y_2}/MPP_{Y_1} = \Delta Y_2/\Delta Y_1$) is less than the slope of the isorevenue line (P_{Y_1}/P_{Y_2}). If $MPP_{Y_2}/MPP_{Y_1} < P_{Y_1}/P_{Y_2}$, then $(MPP_{Y_2})(P_{Y_2}) < (MPP_{Y_1})(P_{Y_1})$ and, by definition, $VMP_{Y_2} < VMP_{Y_1}$. Thus to the left of O the value of lamb given up is less than the value of calf gained.
8. Readers with a mathematical background will recognize that the slope of the transformation curve is actually infinite at point C, requiring a vertical isorevenue curve in order for tangency to occur at C. To simplify the discussion to follow, this bit of mathematical truth will be ignored.
9. Calf prices would have to be extremely high relative to wheat prices in order for the optimum point to fall anywhere close to point C. As with the competitive and complementary relationships, optimum combinations of supplementary products normally involve multiple use as opposed to single use. The only exception to this general rule is the case of antagonistic products discussed below.
10. Inspection of Figures 7.5 and 7.6 reveals that $MRS_{Y_1 Y_2}$ is positive throughout the complementary region and zero in the supplementary region of the above trade-off curves.
11. The snowmobile–skiing example may be unique in terms of its lack of mutual conflict. While viewed as a definite problem by many cross-country skiers, snowmobile users have been known to reply: "We don't mind seeing skiers. Why are they so upset when they see (or hear) us?"

Chapter 8

Discounting—Adjusting Costs and Returns for the Effects of Time

WHY DOES IT COST TO WAIT?

Most investments in range improvements and management practices have a life expectancy in excess of one year. Not only is a current investment expenditure expected to yield increased forage and livestock production over a span of several years, but future costs such as required maintenance of improvements may also be incurred. These various amounts of costs and returns cannot be directly compared. Instead all future changes in revenue and expenditure must be converted to a common point in time. To see why this is true, let us begin with a simple example.

Suppose I offer you your choice of $1 today or $1 one year from now. You undoubtedly will choose the present dollar.[1] In response to my question as to why, you may offer one or more of several plausible reasons. You may point out that because of expected inflation, $1 will probably buy less next year than this year. Or you might observe that if I gave you the $1 now you could deposit it in a savings account at interest and at the end of one year you would have an amount in excess of $1. You might also mention the risk you would be taking

DISCOUNTING—ADJUSTING COSTS AND RETURNS

Table 8.1 Amounts Required Next Year to Equal $1 Today

Value today	Cause of decreased value	Amount received next year
$1	Inflation	$1.05
$1	Opportunity Cost	1.04
$1	Risk	1.05
$1	Combined causes	$1.14

in waiting one year for me to make good on my promise of today. For these several reasons then, the $1 you expect to receive in one year is worth considerably less than $1 today. This illustrates your "time preference" for money. The preference for present benefits over future benefits, shared by individual consumers and investors, lending institutions, and by society in general, can be illustrated another way. Table 8.1 shows the amounts of money that you would have to receive one year from today in order to obtain the equivalent of $1 today after being compensated for the effects of inflation, opportunity cost, and risk. Taking these effects individually, if an inflation rate of five percent is expected during the next year, you would require me to pay $1.05 one year from now in order to provide you the equivalent of $1 of current buying power. Similarly, if you have the opportunity to earn four percent real (inflation-free) interest on funds deposited in a savings account, I must pay you $1.04 in one year to give you the equivalent of $1 today. Next suppose your knowledge of me is sufficient to allow you to calculate the amount of risk you would be taking in trusting me to follow through on a 12-month promise. If, in our past dealings, I have failed to keep 1 out of 20 promises, you are taking a five-percent risk, and $1 of present value would require a payment of $1.05 one year from today. To compensate for the combined effects of all three causes of value reduction, $1.14 must be received after one year to provide the value equivalent of $1.00 received today.

Of the three causes of reduced future value, opportunity cost is the most troublesome. The effects of both inflation and risk can be handled separately from discounting calculations[2] but opportunity cost can be dealt with only through discounting. When first introduced to the concept, many students have difficulty seeing opportunity costs as real costs associated with implementing range management projects. Perhaps because they have generally not yet experienced the cash interest costs of using borrowed funds, they have trouble perceiving the opportunity costs associated with investing one's own money. However, it must be remembered that even when we undertake an investment using our own funds, there is an opportunity cost associated with the use of our own capital similar to interest on borrowed money. If we have funds we are not using, another investor will gladly pay interest to obtain their temporary use. Opportunity cost is simply a measure of this foregone opportunity to earn interest on owned capital.

Table 8.2 Illustration of Compound Interest and Derivation of Compounding Formula

Year	Amount owed ($)	Amount owed (formula for V_n)
1	$100 = principal borrowed	V_0
2	$100 (1 + 0.05) = $105 = $100 principal + $5 interest on principal	$V_0(1 + 0.05)$
3	$100 (1 + 0.05)(1 + 0.05) = $110.25 = $100 principal + $5 interest on principal during year 1 + $5 interest on principal during year 2 + $0.25 interest on interest during year 2	$V_0(1 + 0.05)(1 + 0.05)$ $V_0(1 + 0.05)^2$ or $V_0(1 + i)^n$

i is interest rate; n is number of years.

COMPOUNDING

The examples of Table 8.1 demonstrate the required adjustments to sums to be received one year in the future. In range management we are usually dealing with investments that produce returns for periods from 10 to 30 years and the necessary discounting calculations are more complex. In mathematical terms, multiple year project lives require compound interest calculations.

For example, suppose you lend me $100 at five percent interest, which I am to repay in exactly one year. At the end of one year I will owe $100 principal plus $5 interest. However, when the agreed upon day arrives, I relate the sad story of my financial difficulties during the past year and you agree to carry my $100 loan for a second year. My luck improves and on the due date I present you with $110: $100 principal and $10 interest, $5 for each of two years. Does this satisfy my obligation?

No! You remind me that I have not only enjoyed use for two years of the $100 principal originally borrowed, but during the second year of our agreement I have deprived you of the use of the $5 in interest that was actually due at the end of the first year. Thus my debt to you includes principal, interest on principal, and interest on interest or *compound interest*. As shown in Table 8.2, the total amount due is $110.25, comprising $100 principal, $5 interest on principal during the first year of our loan, $5 interest on principal during the second year of our loan and $0.25 second year interest on the $5 first year interest.

The general compounding formula is

$$V_n = V_0(1 + i)^n,$$

DISCOUNTING—ADJUSTING COSTS AND RETURNS

where V_n is the future value of a present sum at the end of n years, V_0 is the present sum, i is the interest rate charged per period,[3] and n is the number of periods over which V_0 is to be compounded. The formula is somewhat cumbersome to use and requires the use of logarithms. Fortunately compounding (future value) and discounting (present value) tables are available that greatly simplify adjustments of costs and returns for time. Future Worth of One values for interest rates from 3 to 10 percent and years from 1 to 50 are listed in Table 8.3. The values listed may be substituted for $(1 + i)^n$ in the above formula. Returning to the above example, the amount owed at the end of two years on a $100 loan at five percent interest compounded annually is

$$V_n = V_0(1 + i)^n = \$100(1.05)^2 = \$100(1.1025) = \$110.25.$$

The Future Worth of One value, 1.1025, found in the two-year row and the five percent column in Table 8.3 is the equivalent of $(1.05)^2$ and may be substituted directly into the future value formula.

The compounding or future value formula is basic to calculating the value to which money deposited at interest today will grow by some future date. This formula is also useful in decisions involving land investment. Suppose you have the opportunity to buy rangeland for $100 per acre. Your banker will lend money at 10 percent nominal interest (the current market rate, which includes real opportunity cost and inflation) and your best alternative investment opportunity also promises to yield a 10-percent nominal return (total dollar return including both real opportunity cost and inflation). Beginning with the simple case where the land produces no net income and ignoring federal income tax advantages of land ownership, what nominal price per acre (total price, including inflation) must you receive after holding the land for 10 years in order to exactly break even? This question is easily answered by establishing the following equation:

$$V_n = V_0(1 + i)^n = \$100(1 + 0.10)^{10}$$
$$= \$100(\text{Future Worth of One for 10 years at 10 percent})$$
$$= \$100(2.5937) = \$259.37 \text{ per acre.}$$

Your conclusion is that if land values can safely be expected to appreciate by slightly more than $2\frac{1}{2}$ times during the next 10 years, you can afford to buy the land today at $100 per acre with money borrowed from the bank at 10 percent nominal interest and still break even on your investment.[4] But would you really undertake an investment that promises to only break even? Equally important, would your banker back this deal? Banks normally loan a maximum of only 70 percent of appraised real estate value. Thus this investment would actually involve a $30 per acre cash down payment and $70 per acre bank loan at 10 percent nominal interest. The real question then, is whether expected land appreciation is sufficient to cover bank interest and opportunity

Table 8.3 Future Worth of One[a]

Years	3%	4%	4½%	5%	5½%	6%	6½%	7%	7½%	8%	9%	10%
1	1.0300	1.0400	1.0450	1.0500	1.0550	1.0600	1.0650	1.0700	1.0750	1.0800	1.0900	1.1000
2	1.0609	1.0816	1.0920	1.1025	1.1130	1.1236	1.1342	1.1449	1.1556	1.1664	1.1881	1.2100
3	1.0927	1.1249	1.1412	1.1576	1.1742	1.1910	1.2079	1.2250	1.2422	1.2597	1.2950	1.3310
4	1.1255	1.1699	1.1925	1.2155	1.2388	1.2625	1.2864	1.3107	1.3354	1.3604	1.4115	1.4641
5	1.1593	1.2167	1.2462	1.2763	1.3069	1.3382	1.3700	1.4025	1.4356	1.4693	1.5386	1.6105
6	1.1941	1.2653	1.3023	1.3401	1.3788	1.4185	1.4591	1.5007	1.5433	1.5868	1.6771	1.7715
7	1.2299	1.3159	1.3609	1.4071	1.4546	1.5036	1.5539	1.6057	1.6590	1.7138	1.8280	1.9487
8	1.2668	1.3686	1.4221	1.4775	1.5346	1.5938	1.6549	1.7181	1.7834	1.8509	1.9925	2.1435
9	1.3048	1.4233	1.4861	1.5513	1.6190	1.6895	1.7625	1.8384	1.9172	1.9990	2.1718	2.3579
10	1.3439	1.4802	1.5530	1.6289	1.7081	1.7908	1.8771	1.9671	2.0610	2.1589	2.3673	2.5937
11	1.3842	1.5395	1.6229	1.7103	1.8020	1.8983	1.9991	2.1048	2.2156	2.3316	2.5804	2.8531
12	1.4258	1.6010	1.6959	1.7959	1.9012	2.0122	2.1290	2.2521	2.3817	2.5181	2.8126	3.1384
13	1.4685	1.6651	1.7722	1.8856	2.0057	2.1329	2.2674	2.4098	2.5604	2.7196	3.0658	3.4522
14	1.5126	1.7317	1.8519	1.9799	2.1160	2.2609	2.4148	2.5785	2.7524	2.9371	3.3417	3.7974
15	1.5580	1.8009	1.9353	2.0789	2.2324	2.3966	2.5718	2.7590	2.9588	3.1721	3.6424	4.1772
16	1.6047	1.8730	2.0224	2.1829	2.3552	2.5404	2.7390	2.9521	3.1807	3.4259	3.9703	4.5949
17	1.6528	1.9479	2.1134	2.2920	2.4848	2.6928	2.9170	3.1588	3.4193	3.7000	4.3276	5.0544
18	1.7024	2.0258	2.2085	2.4066	2.6214	2.8543	3.1066	3.3799	3.6758	3.9960	4.7171	5.5599
19	1.7535	2.1068	2.3079	2.5269	2.7656	3.0256	3.3085	3.6165	3.9514	4.3157	5.1416	6.1159
20	1.8061	2.1911	2.4117	2.6533	2.9177	3.2071	3.5236	3.8696	4.2478	4.6609	5.6044	6.7274
21	1.8603	2.2788	2.5202	2.7860	3.0782	3.3996	3.7526	4.1405	4.5664	5.0338	6.1088	7.4002
22	1.9161	2.3699	2.6337	2.9253	3.2475	3.6035	3.9966	4.4304	4.9089	5.4365	6.6586	8.1402
23	1.9736	2.4647	2.7522	3.0715	3.4261	3.8197	4.2563	4.7405	5.2770	5.8714	7.2578	8.9543
24	2.0328	2.5633	2.8760	3.2251	3.6145	4.0489	4.5330	5.0723	5.6728	6.3411	7.9110	9.8497
25	2.0938	2.6658	3.0054	3.3864	3.8133	4.2919	4.8276	5.4274	6.0983	6.8484	8.6230	10.8347

26	2.1566	2.7725	3.1407	3.5557	4.0231	4.5494	5.1414	5.8073	6.5557	7.3963	9.3991	11.9181
27	2.2213	2.8834	3.2820	3.7335	4.2444	4.8223	5.4756	6.2138	7.0473	7.9880	10.2450	13.1099
28	2.2879	2.9987	3.4297	3.9201	4.4778	5.1117	5.8316	6.6488	7.5759	8.6271	11.1671	14.4209
29	2.3566	3.1187	3.5840	4.1161	4.7241	5.4184	6.2106	7.1142	8.1441	9.3172	12.1721	15.8630
30	2.4273	3.2434	3.7453	4.3219	4.9839	5.7435	6.6143	7.6122	8.7549	10.0626	13.2676	17.4494
31	2.5001	3.3731	3.9139	4.5380	5.2580	6.0881	7.0442	8.1451	9.4115	10.8676	14.4617	19.1943
32	2.5751	3.5081	4.0900	4.7649	5.5472	6.4534	7.5021	8.7152	10.1174	11.7370	15.7633	21.1137
33	2.6523	3.6484	4.2740	5.0032	5.8523	6.8406	7.9898	9.3253	10.8762	12.6760	17.1820	23.2251
34	2.7319	3.7943	4.4664	5.2533	6.1742	7.2510	8.5091	9.9781	11.6919	13.6901	18.7284	25.5476
35	2.8139	3.9461	4.6673	5.5160	6.5138	7.6861	9.0622	10.6765	12.5688	14.7853	20.4139	28.1024
36	2.8983	4.1039	4.8774	5.7918	6.8720	8.1473	9.6513	11.4239	13.5115	15.9681	22.2512	30.9126
37	2.9852	4.2681	5.0969	6.0814	7.2500	8.6361	10.2786	12.2236	14.5249	17.2456	24.2538	34.0039
38	3.0748	4.4388	5.3262	6.3855	7.6488	9.1543	10.9467	13.0792	15.6142	18.6252	26.4366	37.4043
39	3.1670	4.6164	5.5659	6.7048	8.0694	9.7035	11.6582	13.9948	16.7853	20.1152	28.8159	41.1447
40	3.2620	4.8010	5.8164	7.0400	8.5133	10.2857	12.4160	14.9744	18.0442	21.7245	31.4094	45.2592
41	3.3599	4.9931	6.0781	7.3920	8.9815	10.9029	13.2231	16.0226	19.3975	23.4624	34.2362	49.7851
42	3.4607	5.1928	6.3516	7.7616	9.4755	11.5570	14.0826	17.1442	20.8523	25.3394	37.3175	54.7636
43	3.5645	5.4005	6.6374	8.1497	9.9966	12.2505	14.9979	18.3443	22.4163	27.3666	40.6761	60.2400
44	3.6715	5.6165	6.9361	8.5571	10.5464	12.9855	15.9728	19.6284	24.0975	29.5559	44.3369	66.2640
45	3.7816	5.8412	7.2484	8.9850	11.1265	13.7646	17.0110	21.0024	25.9048	31.9204	48.3272	72.8904
46	3.8950	6.0748	7.5744	9.4343	11.7385	14.5905	18.1168	22.4726	27.8477	34.4740	52.6767	80.1795
47	4.0119	6.3178	7.9153	9.9060	12.3841	15.4659	19.2944	24.0457	29.9362	37.2320	57.4176	88.1974
48	4.1323	6.5705	8.2715	10.4013	13.0652	16.3939	20.5485	25.7289	32.1815	40.2105	62.5852	97.0172
49	4.2562	6.8333	8.6437	10.9213	13.7838	17.3775	21.8842	27.5299	34.5951	43.4274	68.2179	106.7189
50	4.3839	7.1067	9.0326	11.4674	14.5419	18.4202	23.3066	29.4570	37.1897	46.9016	74.3575	117.3908

[a]Future value of $1 deposited at beginning of current year.

cost on the use of your own down payment. The projected selling price of $259.37 at the end of 10 years provides a 10 percent compounded return on your total $100 investment. Since the nominal bank interest rate on the $70 loan and the nominal rate of return on the best alternative use of your own $30 are both 10 percent, the projected selling price exactly covers the full costs of the investment. The 10 percent compound rate of growth of the $70 pays bank interest and principal. Your own $30 also gains at 10 percent per year, compounded, which provides for a 10 percent return on your capital as well as full return of your down payment at the time of sale. In answer to the question, as a rational investor you would buy the land.[5]

DISCOUNTING

In the preceding section, a present sum was converted to a future sum using the concept of compounding. *Discounting* is the mathematical opposite of compounding and provides a means of calculating the present value of a future sum. The general discounting formula, which can be derived from the compounding formula by simply solving for V_0, is

$$V_0 = \frac{V_n}{(1+i)^n} = V_n(1+i)^{-n},$$

where V_0, V_n, i, and n are as previously defined. This equation allows the calculation of the present value of an amount of money to be received at some point in the future. Present Worth of One values (Table 8.4) may be substituted for $(1+i)^{-n}$.

Looking again at the preceding rangeland investment decision, suppose you can obtain a bank loan at 10 percent nominal interest and your expected rate of return on your best alternative investment is also a nominal 10 percent. If the price of land is expected to double over the next 10 years, can you currently afford to buy land at $100 per acre? This question may be stated in equation form as

$$V_0 = V_n(1+i)^{-n} = \$200(1 + 0.10)^{-10} = \$200(0.3855),$$

where 0.3855 is the Present Worth of One value for 10 years and 10 percent from Table 8.4. The solution is $V_0 = \$77.10$ and you conclude that you cannot afford to pay $100 now for land you expect to sell for $200 at the end of 10 years.

Basing our investment decisions on the solution of V_0 works well when we know both the rate of interest charged for borrowed money and the opportunity cost rate. However, in some cases we do not know either rate. The loan rates charged by banks vary with the financial status, experience, and repayment record of the individual borrower. Because of differences in capital

DISCOUNTING—ADJUSTING COSTS AND RETURNS

position, general knowledge, and degree of risk aversion, there is also considerable variation in rates of return on alternative investment opportunities faced by investors. Therefore a somewhat different approach is required when we are making land investment calculations for investors in general. Continuing with the same example, suppose current land price is $100 per acre, price is expected to double during the next 10 years, and we want to provide an interpretation of this growth in land value applicable to any potential investor. Our interpretation will be based on the rate of return generated by the doubling in land value during the 10 year period and the discounting formula again forms the basis of our calculations. Substituting available information into $V_0 = V_n(1 + i)^{-n}$ we obtain

$$\$100 = \$200(1 + i)^{-10} \quad \text{or}$$
$$\$100 = \$200 \quad \text{(Present Worth of One for 10 years and } i \text{ percent).}$$

Dividing both sides by $200 gives 0.5000, the Present Worth of One value for 10 years in Table 8.4 that corresponds to the unknown interest rate, i, that we are seeking. On the 10-year row of Table 8.4 are Present Worth of One values of 0.5083 for 7 percent and 0.4852 for $7\frac{1}{2}$ percent. These two rates bracket i, and closer examination reveals that the doubling in land value generates a rate of return slightly over 7 percent. Based on this 7 percent rate of return our general recommendation concerning this investment opportunity can be stated as follows: If an investor can borrow funds for 7 percent and if there is no alternative investment opportunity that promises to yield more than 7 percent, investment in this land is feasible. However, if either loan rate or opportunity cost faced by the individual investor exceeds 7 percent, he cannot afford to invest in this land. Once this information has been presented, any investor can supply his own borrowing and opportunity cost rates and make his own rational decision.

FLOW OF COSTS AND RETURNS

All of the above compounding and discounting calculations were based on *stocks* (single sums of money received or paid out at single points in time). In range management we work more often with *flows* (several sums of money received or paid out over an extended period of time). Range improvements often involve an initial investment (a stock) with net returns accruing as a flow over time. Stocks and flows can be compared only after being converted to a common point in time. Discounting is again the technique used for conversion.

The general formula for discounting uniform flows (annuities) is:

$$V_0 = R \frac{[1 - (1 + i)^{-n}]}{i},$$

Table 8.4 Present Worth of One[a]

Years	3%	4%	4½%	5%	5½%	6%	6½%	7%	7½%	8%	9%	10%	11%	12%	13%	14%	15%
1	0.9709	0.9615	0.9569	0.9524	0.9479	0.9434	0.9390	0.9346	0.9302	0.9259	0.9174	0.9091	0.9009	0.8929	0.8850	0.8772	0.8696
2	0.9426	0.9246	0.9157	0.9070	0.8985	0.8900	0.8817	0.8734	0.8653	0.8573	0.8417	0.8264	0.8116	0.7972	0.7831	0.7695	0.7561
3	0.9151	0.8890	0.8763	0.8638	0.8516	0.8396	0.8278	0.8163	0.8050	0.7938	0.7722	0.7513	0.7312	0.7118	0.6930	0.6750	0.6575
4	0.8885	0.8548	0.8386	0.8227	0.8072	0.7921	0.7773	0.7629	0.7488	0.7350	0.7084	0.6830	0.6587	0.6355	0.6133	0.5921	0.5718
5	0.8626	0.8219	0.8025	0.7835	0.7651	0.7473	0.7299	0.7130	0.6966	0.6806	0.6499	0.6209	0.5935	0.5674	0.5428	0.5194	0.4972
6	0.8375	0.7903	0.7679	0.7462	0.7252	0.7050	0.6853	0.6663	0.6480	0.6302	0.5963	0.5645	0.5346	0.5066	0.4803	0.4556	0.4323
7	0.8131	0.7599	0.7348	0.7107	0.6874	0.6651	0.6435	0.6227	0.6027	0.5835	0.5470	0.5132	0.4816	0.4523	0.4250	0.3996	0.3759
8	0.7894	0.7307	0.7032	0.6768	0.6516	0.6274	0.6042	0.5820	0.5607	0.5403	0.5019	0.4665	0.4339	0.4039	0.3762	0.3506	0.3269
9	0.7664	0.7026	0.6729	0.6446	0.6176	0.5919	0.5673	0.5439	0.5216	0.5002	0.4604	0.4241	0.3909	0.3606	0.3329	0.3075	0.2843
10	0.7441	0.6756	0.6439	0.6139	0.5854	0.5584	0.5327	0.5083	0.4852	0.4632	0.4224	0.3855	0.3522	0.3220	0.2946	0.2697	0.2472
11	0.7224	0.6496	0.6162	0.5847	0.5549	0.5268	0.5002	0.4751	0.4514	0.4289	0.3875	0.3505	0.3173	0.2875	0.2607	0.2366	0.2149
12	0.7014	0.6246	0.5897	0.5568	0.5260	0.4970	0.4697	0.4440	0.4199	0.3971	0.3555	0.3186	0.2858	0.2567	0.2307	0.2075	0.1869
13	0.6810	0.6006	0.5643	0.5303	0.4986	0.4688	0.4410	0.4150	0.3906	0.3677	0.3262	0.2897	0.2575	0.2292	0.2042	0.1821	0.1625
14	0.6611	0.5775	0.5400	0.5051	0.4726	0.4423	0.4141	0.3878	0.3633	0.3405	0.2992	0.2633	0.2320	0.2046	0.1807	0.1597	0.1413
15	0.6419	0.5553	0.5167	0.4810	0.4479	0.4173	0.3888	0.3624	0.3380	0.3152	0.2745	0.2394	0.2090	0.1827	0.1599	0.1401	0.1229
16	0.6232	0.5339	0.4945	0.4581	0.4246	0.3936	0.3651	0.3387	0.3144	0.2919	0.2519	0.2176	0.1883	0.1631	0.1415	0.1229	0.1069
17	0.6050	0.5134	0.4732	0.4363	0.4024	0.3714	0.3428	0.3166	0.2924	0.2703	0.2311	0.1978	0.1696	0.1456	0.1262	0.1078	0.0929
18	0.5874	0.4936	0.4528	0.4155	0.3815	0.3503	0.3219	0.2959	0.2720	0.2502	0.2120	0.1799	0.1528	0.1300	0.1108	0.0946	0.0808
19	0.5703	0.4746	0.4333	0.3957	0.3616	0.3305	0.3022	0.2765	0.2531	0.2317	0.1945	0.1635	0.1377	0.1161	0.0981	0.0829	0.0703
20	0.5537	0.4564	0.4146	0.3769	0.3427	0.3118	0.2838	0.2584	0.2354	0.2145	0.1784	0.1486	0.1240	0.1037	0.0868	0.0728	0.0611
21	0.5375	0.4388	0.3968	0.3589	0.3249	0.2942	0.2665	0.2415	0.2190	0.1987	0.1637	0.1351	0.1117	0.0925	0.0758	0.0638	0.0531
22	0.5219	0.4220	0.3797	0.3418	0.3079	0.2775	0.2502	0.2257	0.2037	0.1839	0.1502	0.1228	0.1007	0.0826	0.0680	0.0560	0.0462
23	0.5067	0.4057	0.3633	0.3256	0.2919	0.2618	0.2349	0.2109	0.1895	0.1703	0.1378	0.1117	0.0907	0.0738	0.0601	0.0491	0.0402
24	0.4919	0.3901	0.3477	0.3101	0.2766	0.2470	0.2206	0.1971	0.1763	0.1577	0.1264	0.1015	0.0817	0.0659	0.0532	0.0431	0.0349
25	0.4776	0.3751	0.3327	0.2953	0.2622	0.2330	0.2071	0.1842	0.1640	0.1460	0.1160	0.0923	0.0736	0.0588	0.0471	0.0378	0.0304

26	0.4637	0.3607	0.3184	0.2812	0.2486	0.2198	0.1945	0.1722	0.1525	0.1352	0.1064	0.0839	0.0663	0.0525	0.0417	0.0331	0.0264
27	0.4502	0.3468	0.3047	0.2678	0.2356	0.2074	0.1826	0.1609	0.1419	0.1252	0.0976	0.0763	0.0597	0.0469	0.0369	0.0291	0.0230
28	0.4371	0.3335	0.2916	0.2551	0.2233	0.1956	0.1715	0.1504	0.1320	0.1159	0.0895	0.0693	0.0538	0.0419	0.0326	0.0255	0.0200
29	0.4243	0.3207	0.2790	0.2429	0.2117	0.1846	0.1610	0.1406	0.1228	0.1073	0.0822	0.0630	0.0485	0.0374	0.0289	0.0224	0.0174
30	0.4120	0.3083	0.2670	0.2314	0.2006	0.1741	0.1512	0.1314	0.1142	0.0994	0.0754	0.0573	0.0437	0.0334	0.0256	0.0196	0.0151
31	0.4000	0.2965	0.2555	0.2204	0.1902	0.1643	0.1420	0.1228	0.1063	0.0920	0.0691	0.0521	0.0394	0.0298	0.0226	0.0172	0.0131
32	0.3883	0.2851	0.2445	0.2099	0.1803	0.1550	0.1333	0.1147	0.0988	0.0852	0.0634	0.0474	0.0354	0.0266	0.0200	0.0151	0.0114
33	0.3770	0.2741	0.2340	0.1999	0.1709	0.1462	0.1251	0.1072	0.0919	0.0789	0.0582	0.0431	0.0319	0.0238	0.0177	0.0132	0.0099
34	0.3660	0.2636	0.2239	0.1904	0.1620	0.1379	0.1175	0.1002	0.0855	0.0730	0.0534	0.0391	0.0288	0.0212	0.0157	0.0116	0.0086
35	0.3554	0.2534	0.2142	0.1813	0.1535	0.1301	0.1103	0.0937	0.0796	0.0676	0.0490	0.0356	0.0259	0.0189	0.0139	0.0102	0.0075
36	0.3450	0.2437	0.2050	0.1727	0.1455	0.1227	0.1036	0.0875	0.0740	0.0626	0.0449	0.0323	0.0234	0.0169	0.0123	0.0089	0.0065
37	0.3350	0.2343	0.1962	0.1644	0.1379	0.1158	0.0973	0.0818	0.0688	0.0580	0.0412	0.0294	0.0210	0.0151	0.0109	0.0078	0.0057
38	0.3252	0.2253	0.1878	0.1566	0.1307	0.1092	0.0914	0.0765	0.0640	0.0537	0.0378	0.0267	0.0189	0.0135	0.0096	0.0069	0.0049
39	0.3158	0.2166	0.1797	0.1491	0.1239	0.1031	0.0858	0.0715	0.0596	0.0497	0.0347	0.0243	0.0171	0.0120	0.0085	0.0060	0.0043
40	0.3066	0.2083	0.1719	0.1420	0.1175	0.0972	0.0805	0.0668	0.0554	0.0460	0.0318	0.0221	0.0154	0.0107	0.0075	0.0053	0.0037
41	0.2976	0.2003	0.1645	0.1353	0.1113	0.0917	0.0756	0.0624	0.0515	0.0426	0.0292	0.0201	0.0139	0.0096	0.0067	0.0046	0.0032
42	0.2890	0.1926	0.1574	0.1288	0.1055	0.0865	0.0710	0.0583	0.0480	0.0395	0.0268	0.0183	0.0125	0.0086	0.0059	0.0041	0.0028
43	0.2805	0.1852	0.1507	0.1227	0.1000	0.0816	0.0667	0.0545	0.0446	0.0365	0.0246	0.0166	0.0112	0.0076	0.0062	0.0036	0.0025
44	0.2724	0.1780	0.1442	0.1169	0.0948	0.0770	0.0626	0.0509	0.0415	0.0338	0.0225	0.0151	0.0101	0.0068	0.0046	0.0031	0.0021
45	0.2644	0.1712	0.1380	0.1113	0.0899	0.0726	0.0588	0.0476	0.0386	0.0313	0.0207	0.0137	0.0091	0.0061	0.0041	0.0027	0.0019
46	0.2567	0.1646	0.1320	0.1060	0.0852	0.0685	0.0552	0.0445	0.0359	0.0290	0.0190	0.0125	0.0082	0.0054	0.0036	0.0024	0.0016
47	0.2493	0.1583	0.1263	0.1009	0.0807	0.0647	0.0518	0.0416	0.0334	0.0269	0.0174	0.0113	0.0074	0.0049	0.0032	0.0021	0.0114
48	0.2420	0.1522	0.1209	0.0961	0.0765	0.0610	0.0487	0.0389	0.0311	0.0249	0.0160	0.0103	0.0067	0.0043	0.0028	0.0019	0.0012
49	0.2350	0.1463	0.1157	0.0916	0.0725	0.0575	0.0457	0.0363	0.0289	0.0230	0.0147	0.0094	0.0060	0.0039	0.0025	0.0016	0.0011
50	0.2281	0.1407	0.1107	0.0872	0.0688	0.0543	0.0429	0.0339	0.0269	0.0213	0.0134	0.0085	0.0054	0.0035	0.0022	0.0014	0.0009

[a] Present value of $1 received at end of future year.

Table 8.5 Present Worth of One Per Period[a]

Years	3%	4%	4½%	5%	5½%	6%	6½%	7%	7½%	8%	9%	10%	11%	12%	13%	14%
1	0.971	0.961	0.957	0.952	0.948	0.943	0.939	0.935	0.930	0.926	0.917	0.909	0.901	0.893	0.885	0.877
2	1.913	1.886	1.873	1.859	1.846	1.833	1.821	1.808	1.796	1.783	1.759	1.736	1.713	1.690	1.668	1.647
3	2.829	2.775	2.749	2.723	2.698	2.673	2.648	2.624	2.600	2.577	2.531	2.487	2.444	2.402	2.361	2.322
4	3.717	3.630	3.587	3.546	3.505	3.465	3.426	3.387	3.349	3.312	3.240	3.170	3.102	3.037	2.974	2.914
5	4.580	4.452	4.390	4.329	4.270	4.212	4.156	4.100	4.046	3.993	3.890	3.791	3.696	3.605	3.517	3.433
6	5.417	5.242	5.158	5.076	4.996	4.917	4.841	4.766	4.694	4.623	4.486	4.355	4.231	4.111	3.998	3.889
7	6.230	6.002	5.893	5.786	5.683	5.582	5.485	5.389	5.297	5.206	5.033	4.868	4.712	4.564	4.423	4.288
8	7.020	6.733	6.596	6.463	6.334	6.210	6.089	5.971	5.857	5.747	5.535	5.335	5.146	4.968	4.799	4.639
9	7.786	7.435	7.269	7.108	6.952	6.802	6.656	6.515	6.379	6.247	5.995	5.759	5.537	5.328	5.132	4.946
10	8.530	8.111	7.913	7.722	7.538	7.360	7.189	7.024	6.864	6.710	6.418	6.145	5.889	5.650	5.426	5.216
11	9.253	8.760	8.529	8.306	8.093	7.887	7.689	7.499	7.315	7.139	6.805	6.495	6.206	5.938	5.687	5.453
12	9.954	9.385	9.118	8.863	8.618	8.384	8.159	7.943	7.735	7.536	7.161	6.814	6.492	6.194	5.918	5.660
13	10.635	9.986	9.683	9.394	9.117	8.853	8.600	8.358	8.126	7.904	7.487	7.103	6.750	6.424	6.122	5.842
14	11.296	10.563	10.223	9.899	9.590	9.295	9.014	8.745	8.489	8.244	7.786	7.367	6.982	6.628	6.302	6.002
15	11.938	11.118	10.739	10.380	10.038	9.712	9.403	9.108	8.827	8.559	8.061	7.606	7.191	6.811	6.462	6.142
16	12.561	11.652	11.234	10.838	10.462	10.106	9.768	9.447	9.142	8.851	8.313	7.824	7.379	6.974	6.604	6.265
17	13.166	12.166	11.707	11.274	10.865	10.477	10.110	9.763	9.434	9.122	8.544	8.022	7.549	7.120	6.729	6.373
18	13.753	12.659	12.160	11.690	11.246	10.828	10.432	10.059	9.706	9.372	8.756	8.201	7.702	7.250	6.840	6.467
19	14.324	13.134	12.593	12.085	11.608	11.158	10.735	10.336	9.959	9.604	8.950	8.365	7.839	7.366	6.938	6.550
20	14.877	13.590	13.008	12.462	11.950	11.470	11.019	10.594	10.194	9.818	9.128	8.514	7.963	7.469	7.025	6.623
21	15.415	14.029	13.405	12.821	12.275	11.764	11.285	10.835	10.413	10.017	9.292	8.649	8.075	7.562	7.102	6.687
22	15.937	14.451	13.784	13.163	12.583	12.042	11.535	11.061	10.617	10.201	9.442	8.772	8.176	7.645	7.170	6.743
23	16.444	14.857	14.148	13.489	12.875	12.303	11.770	11.272	10.807	10.371	9.580	8.883	8.266	7.718	7.230	6.792
24	16.935	15.247	14.495	13.799	13.152	12.550	11.991	11.469	10.983	10.529	9.707	8.985	8.348	7.784	7.283	6.835
25	17.413	15.622	14.828	14.094	13.414	12.783	12.198	11.654	11.147	10.675	9.823	9.077	8.422	7.843	7.330	6.873

26	17.877	15.983	15.147	14.375	13.662	13.003	12.392	11.826	11.299	10.810	9.929	9.161	8.488	7.896	7.372	6.906
27	18.327	16.330	15.451	14.643	13.898	13.210	12.575	11.987	11.441	10.935	10.026	9.237	8.548	7.943	7.409	6.935
28	18.764	16.663	15.743	14.898	14.121	13.406	12.746	12.137	11.573	11.051	10.116	9.307	8.602	7.984	7.441	6.961
29	19.188	16.984	16.022	15.141	14.333	13.591	12.907	12.278	11.696	11.158	10.198	9.370	8.650	8.022	7.470	6.983
30	19.600	17.292	16.289	15.372	14.534	13.765	13.059	12.409	11.810	11.258	10.274	9.427	8.694	8.055	7.496	7.003
31	20.000	17.588	16.544	15.593	14.724	13.929	13.201	12.532	11.917	11.350	10.343	9.479	8.733	8.085	7.518	7.020
32	20.389	17.874	16.789	15.803	14.904	14.084	13.334	12.647	12.015	11.435	10.406	9.526	8.769	8.112	7.538	7.035
33	20.766	18.148	17.023	16.002	15.075	14.230	13.459	12.754	12.107	11.514	10.464	9.569	8.801	8.135	7.556	7.048
34	21.132	18.411	17.247	16.193	15.237	14.368	13.577	12.854	12.193	11.587	10.518	9.609	8.829	8.157	7.572	7.060
35	21.487	18.665	17.461	16.374	15.390	14.498	13.687	12.948	12.272	11.655	10.567	9.644	8.855	8.176	7.586	7.070
36	21.832	18.908	17.666	16.547	15.536	14.621	13.791	13.035	12.347	11.717	10.612	9.676	8.879	8.193	7.598	7.079
37	22.167	19.143	17.862	16.711	15.674	14.737	13.888	13.117	12.415	11.775	10.653	9.706	8.900	8.207	7.609	7.087
38	22.492	19.368	18.050	16.868	15.805	14.846	13.979	13.193	12.479	11.829	10.691	9.733	8.919	8.221	7.618	7.094
39	22.808	19.584	18.230	17.017	15.929	14.949	14.065	13.265	12.539	11.879	10.726	9.757	8.936	8.233	7.627	7.100
40	23.115	19.793	18.401	17.159	16.046	15.046	14.145	13.332	12.594	11.925	10.757	9.779	8.951	8.244	7.634	7.105
41	23.412	19.993	18.566	17.294	16.157	15.138	14.221	13.394	12.646	11.967	10.786	9.799	8.965	8.253	7.641	7.110
42	23.701	20.186	18.724	17.423	16.263	15.224	14.292	13.452	12.694	12.007	10.813	9.817	8.977	8.262	7.647	7.114
43	23.982	20.371	18.874	17.546	16.363	15.306	14.359	13.507	12.738	12.043	10.838	9.834	8.989	8.270	7.652	7.117
44	24.254	20.549	19.018	17.663	16.458	15.383	14.421	13.558	12.780	12.077	10.861	9.849	8.999	8.276	7.657	7.120
45	24.519	20.720	19.156	17.774	16.548	15.456	14.480	13.605	12.819	12.108	10.881	9.863	9.008	8.283	7.661	7.123
46	24.775	20.885	19.288	17.880	16.633	15.524	14.535	13.650	12.855	12.137	10.900	9.875	9.016	8.288	7.664	7.126
47	25.025	21.043	19.415	17.981	16.714	15.589	14.587	13.692	12.888	12.164	10.918	9.887	9.024	8.293	7.668	7.128
48	25.267	21.195	19.536	18.077	16.790	15.650	14.636	13.730	12.919	12.189	10.933	9.897	9.030	8.297	7.670	7.130
49	25.502	21.341	19.651	18.169	16.863	15.708	14.682	13.767	12.948	12.212	10.948	9.906	9.036	8.301	7.673	7.131
50	25.730	21.482	19.762	18.256	16.931	15.762	14.724	13.801	12.975	12.233	10.962	9.915	9.042	8.305	7.675	7.133

^a Present value of $1 payable annually at end of future years.

where V_0 is the present value of a future flow, R is the uniform net annual return received or net annual cost paid out, i is the interest rate, and n is the number of years that R is received or dispersed. This formula is more complex than previous formulas. Fortunately Present Worth of One Per Period values (Table 8.5) may be substituted directly for $[1 - (1 + i)^{-n}]/i$ in the formula.

Illustration is provided by a simple example: Suppose a rancher has 100 acres of depleted private rangeland with sufficiently deep soil and adequate precipitation for artificial reseeding. The cost of reseeding (a present stock) is $18 per acre including seedbed preparation, seed, drilling, and grazing deferment. The present carrying capacity is 6 acres per AUM. Reseeding is expected to double carrying capacity and the projected life of the seeding is 20 years. The rancher currently leases private spring range from a neighbor at $7.50 per AUM. Investment capital is available from the bank at four percent real[6] interest. Is the proposed reseeding a sound range improvement investment? Our answer will be based on the formula for discounting flows of income.

At 6 acres per AUM, current carrying capacity of the 100-acre tract is about 17 AUM and the reseeding promises to provide an additional 17 AUM annually. Since the reseeding will allow the rancher to lease 17 AUM less from his neighbor, the increased AUM can be valued at $7.50, yielding $127.50 added annual net return. Substituting this information into the flow discounting formula along with the appropriate value from Table 8.5 we obtain: $V_0 = R[1 - (1 + i)^{-n}]/i = 127.50[1 - (1.04)^{-20}]/0.04 = 127.50(13.59) = \$1732.73$. Profit accruing to the reseeding is $1732.73 - \$1800$, or a loss of $67.27. Our conclusion is that the rancher cannot afford to invest $1800 in a reseeding that promises to yield a flow of net returns with a present value of only $1732.73.[7]

The preceding approach works well for the situation where both the interest rate charged for borrowed funds and the opportunity cost of owned capital are known. For the case where our conclusions and recommendations are intended for a ranch owner or range manager for whom we have no information concerning either opportunity cost or borrowing rate, a slightly different approach is called for. The rate of return generated on investment in the reseeding, the *internal rate of return* (IRR), is calculated and presented for comparison with the individual investor's borrowing and opportunity cost. Calculation of IRR involves substitution of the $1800 initial investment for I in the formula for the present value of a future flow:

$$I = R\frac{[1-(1+i)^{-n}]}{i}.$$

Substituting our reseeding data we have

$$\$1800 = \$127.50\frac{[1-(1+i)^{-20}]}{i} \quad \text{or}$$

$$\frac{\$1800}{\$127.50} = 14.118 = \frac{[1-(1+i)^{-20}]}{i}.$$

The right hand side of the preceding equation is equivalent to the Present Worth of One Per Period in Table 8.5 corresponding to 20 years and i, the IRR generated by the reseeding project. Inspection of Table 8.5 reveals that the calculated Present Worth value, 14.118, corresponds to an i (IRR) between three and four percent. The IRR is the interest rate that discounts or "forces" a future stream of net returns to just equal the investment required to produce the flow of returns (Nielsen, 1967). Our calculated IRR of just over $3\frac{1}{2}$ percent can be interpreted as follows: If range improvement funds can be borrowed at less than $3\frac{1}{2}$ percent real interest and if no other investment opportunity is available that promises to yield a real return greater than $3\frac{1}{2}$ percent, then the reseeding project is economically feasible.

CAPITALIZATION

Real estate appraisers and lending institutions often base their estimates of current real estate values on a modified procedure for discounting future flows. *Capitalization* is a technique used to calculate the present value of a future perpetual flow of annual net income. Suppose an investor is considering the purchase of a ranch property that promises to yield an annual real net[8] income to land and improvements of $7200 and that investment capital is available from his bank at a real borrowing rate of four percent. These numbers may be substituted into the discounting formula of the previous section as follows:

$$V_0 = \$7200 \frac{[1 - (1.04)^{-n}]}{0.04}.$$

Land is a perpetual asset with an infinite life, and as n approaches infinity, the term $(1.04)^{-n}$ becomes zero. Thus the capitalization formula is reduced to

$$V_0 = \frac{\$7200(1 - 0)}{0.04} = \frac{\$7200}{0.04} = \$180{,}000.$$

Based on his real borrowing cost (or real opportunity cost, if it exceeds borrowing cost) of 4 percent, the ranch is worth $180,000 to the investor since the $7200 real net return is sufficient to pay 4 percent real interest on both a 30 percent down payment of $54,000 and the $126,000 of borrowed capital.[9] The borrowing cost and opportunity cost rates are crucial determinants of real estate value. If the rate were only three percent, the ranch would be worth $240,000, while if the same investor had to pay five percent real interest for borrowed capital, he could only afford to pay $144,000 for the same ranch.

Ranch appraisers often calculate a "market" capitalization rate, based on recent ranch sales, by dividing net return to land and buildings by sale price. Such calculated rates are usually only two to three percent compared with real borrowing and real opportunity cost rates of around four percent. This discrepancy sometimes causes observers to wonder why ranch prices are so high and why they continue to increase in the face of such low net returns.

However, high land prices are a self-perpetuating phenomenon. Land prices continue to increase because ranch investors think land values will go up.

A further application of discounting may help explain the mechanism underlying increasing rangeland prices. Based on a real net ranch income of $7200 annually and the four percent real interest rate charged for borrowed funds, the investor of our example can afford to pay only $180,000 for the ranch in question. However, the $7200 annual net ranch income is only one of two sources of expected return. The investor also expects a return in the form of increased land prices. Even if the investor fully intends to retain ownership of the ranch for many years, he knows that the ranch could be sold at a profit almost anytime during his planned ownership. Land appreciation as a source of future return is at least as important as annual net ranch income.

Suppose our ranch investor plans to hold the property for 10 years and then sell the ranch at its appreciated price. Based on this plan, the current value of the ranch to this particular investor is the present value of the $7200 annual income flow plus the present value of the projected sale price. Referring to Table 8.5 the present value (at four percent real interest) of the 10-year stream of net ranch income is

$$V_0 = \$7200(8.111) = \$58,399.$$

Suppose further that the current asking price for the ranch is its true productive value of $180,000. If the ranch were *not* expected to increase in value during the next 10 years, the present value (at four percent real interest) of the projected selling price of $180,000 in 10 years (see Table 8.4) would be

$$V_0 = \$180,000(0.6756) = \$121,608.$$

The sum of these two present values, $180,007, is identical (except for rounding) to the capitalized value of the $7200 perpetual flow of net ranch income.

However, an increase in land prices *is* expected to occur during the next 10 years. Suppose the expected general inflation rate for this period is six percent. If the investor expects land prices to increase somewhat faster than general inflation, let us say at seven percent annually, the projected nominal sale price (applying the appropriate value from Table 8.3) is

$$V_n = \$180,000(1.9671) = \$354,078.$$

Now discounting at a 10 percent nominal borrowing rate (4 percent real opportunity cost plus 6 percent expected inflation), we calculate the present value of the projected sale price to be (Table 8.4)

$$V_0 = \$354,078(0.3855) = \$136,497.$$

DISCOUNTING—ADJUSTING COSTS AND RETURNS

Thus the investor facing a nominal interest rate of 10 percent for borrowed funds can afford to pay $136,497 now for the projected future value of the ranch. He can afford to pay an additional $58,399 for the projected 10-year flow of real net ranch income. The total maximum present value of the ranch to the investor, then, is $194,896.[10] This figure can be interpreted as the maximum amount the investor could pay for the $180,000 ranch and still break even. Thus expectations concerning land price increases lead to continued actual price rises despite low net ranch incomes. A second partial explanation of high land prices may be the federal income tax advantages associated with land ownership, especially the range improvement current cost deduction and capital gains provisions (Workman and Hooper, 1975).

SINKING FUND

Another useful application of the compounding concept is known as the *sinking fund*. This "fund" is a conceptual accounting device for gradually setting aside funds over a period of time so that capital required for some anticipated future expenditure will be available. Suppose a rancher owns a tractor purchased recently for $20,000. The tractor is expected to have a useful life of about five years and for accounting purposes we could simply attribute one-fifth of the purchase price, minus salvage, to annual depreciation, the amount of tractor value "used up" annually. Due to inflation, though, the rancher knows that in five years the price of the tractor will be considerably more than $20,000. Also, the funds earmarked annually for purchase of the replacement will be placed in a savings account and will be accumulating five percent interest prior to expenditure.

If inflation is expected to average seven percent over the next five years, Table 8.3 can be used to calculate the price of the replacement tractor:

$$V_n = \$20,000(1.4025) = \$28,050.$$

The general formula for calculating the annual payment that must be deposited at interest in order to have a certain sum available at a specific point in the future is

$$V_a = V_n \left[\frac{(i)}{(1+i)^n - 1} \right] \quad \text{or}$$

$$V_a = \$28,050 \quad \text{(Annual Deposit to Yield One)},$$

where V_a is the necessary amount to be deposited annually at the end of each year, V_n is the amount required at the end of the period, i is the rate of interest paid on money deposited in the sinking fund, and n is the number of years until the fund must be expended. Substituting the Annual Deposit to Yield One value in Table 8.6 corresponding to five years and five percent, we obtain

$$V_a = \$28,050(0.18097) = \$5076.21.$$

Thus instead of $4000 (one-fifth of the current purchase price of the tractor),

Table 8.6 Annual Deposit to Yield One[a]

Years	1/3%	1%	1 1/2%	1 3/4%	2%	2 1/2%	3%	3 1/2%	4%	4 1/2%	5%	5 1/2%	6%	7%
1	1.00000	1.00000	1.00000	1.00000	1.00000	1.00000	1.00000	1.00000	1.00000	1.00000	1.00000	1.00000	1.00000	1.00000
2	0.49917	0.49751	0.49628	0.49566	0.49505	0.49383	0.42961	0.49140	0.49020	0.48900	0.48780	0.48662	0.48544	0.48309
3	0.33222	0.33002	0.32838	0.32757	0.32675	0.32514	0.32353	0.32193	0.32035	0.31877	0.31721	0.31565	0.31411	0.31105
4	0.24875	0.24628	0.24444	0.24353	0.24262	0.24082	0.23903	0.23725	0.23549	0.23374	0.23201	0.23029	0.22859	0.22523
5	0.19867	0.19604	0.19409	0.19312	0.19216	0.19025	0.18835	0.18648	0.18463	0.18279	0.18097	0.17918	0.17740	0.17389
6	0.16528	0.16255	0.16053	0.15952	0.15853	0.15655	0.15460	0.15267	0.15076	0.14888	0.14702	0.14518	0.14336	0.13980
7	0.14143	0.13863	0.13656	0.13553	0.13451	0.13250	0.13051	0.12854	0.12661	0.12470	0.12282	0.12096	0.11913	0.11555
8	0.12355	0.12069	0.11858	0.11754	0.11651	0.11447	0.11246	0.11048	0.10853	0.10661	0.10472	0.10286	0.10104	0.09757
9	0.10964	0.10674	0.10461	0.10356	0.10252	0.10046	0.09843	0.09634	0.09449	0.09257	0.09069	0.08884	0.08702	0.08349
10	0.09851	0.09558	0.09343	0.09238	0.09133	0.08926	0.08723	0.08524	0.08329	0.08138	0.07950	0.07767	0.07587	0.07238
11	0.08940	0.08645	0.08429	0.08323	0.08218	0.08011	0.07808	0.07609	0.07415	0.07225	0.07039	0.06857	0.06679	0.06336
12	0.08182	0.07885	0.07668	0.07561	0.07456	0.07249	0.07046	0.06848	0.06655	0.06467	0.06283	0.06103	0.05928	0.06690
13	0.07539	0.07241	0.07024	0.06917	0.06812	0.06605	0.06403	0.06206	0.06014	0.05828	0.05646	0.05468	0.05296	0.05965
14	0.06989	0.06690	0.06472	0.06366	0.06260	0.06054	0.05853	0.05657	0.05467	0.05282	0.05102	0.04928	0.04758	0.04434
15	0.06512	0.06212	0.05994	0.05888	0.05783	0.05577	0.05377	0.05183	0.04994	0.04811	0.04634	0.04463	0.04296	0.03979
16	0.06095	0.05794	0.05577	0.05470	0.05365	0.05160	0.04961	0.04768	0.04582	0.04402	0.04227	0.04058	0.03895	0.03586
17	0.05727	0.05426	0.05208	0.05102	0.04997	0.04793	0.04595	0.04404	0.04220	0.04042	0.03870	0.03704	0.03544	0.03243
18	0.05400	0.05098	0.04881	0.04774	0.04670	0.04467	0.04271	0.04082	0.03899	0.03724	0.03555	0.03392	0.03236	0.02941
19	0.05107	0.04805	0.04588	0.04482	0.04378	0.04176	0.03981	0.03794	0.03614	0.03441	0.03275	0.03115	0.02962	0.02675
20	0.04843	0.04541	0.04325	0.04219	0.04116	0.03915	0.03722	0.03536	0.03358	0.03188	0.03024	0.02868	0.02718	0.02439
21	0.04605	0.04303	0.04087	0.03981	0.03878	0.03679	0.03487	0.03304	0.03128	0.02960	0.02800	0.02646	0.02500	0.02229
22	0.04388	0.04086	0.03870	0.03766	0.03663	0.03465	0.03275	0.03093	0.02920	0.02755	0.02597	0.02447	0.02305	0.02041
23	0.04190	0.03888	0.03673	0.03569	0.03467	0.03270	0.03081	0.02902	0.02731	0.02568	0.02414	0.02267	0.02128	0.01871
24	0.04009	0.03707	0.03492	0.03389	0.03287	0.03091	0.02905	0.02727	0.02559	0.02399	0.02247	0.02104	0.01968	0.01719
25	0.03842	0.03541	0.03326	0.03223	0.03122	0.02928	0.02743	0.02567	0.02401	0.02244	0.02095	0.01955	0.01823	0.01581
26	0.03688	0.03387	0.03173	0.03070	0.02970	0.02777	0.02594	0.02421	0.02257	0.02102	0.01956	0.01819	0.01690	0.01456
27	0.03546	0.03244	0.03032	0.02929	0.02829	0.02638	0.02456	0.02285	0.02124	0.01972	0.01829	0.01695	0.01570	0.01343
28	0.03413	0.03112	0.02900	0.02798	0.02699	0.02509	0.02329	0.02160	0.02001	0.01852	0.01712	0.01581	0.01459	0.01239
29	0.03290	0.02989	0.02778	0.02676	0.02578	0.02389	0.02211	0.02045	0.01888	0.01741	0.01605	0.01477	0.01358	0.01145
30	0.03175	0.02875	0.02664	0.02563	0.02465	0.02278	0.02102	0.01937	0.01783	0.01639	0.01505	0.01381	0.01265	0.01059

[a]End of year annual deposit required to yield $1 in the future.

the rancher must deposit $5076.21 each year in order to have sufficient funds, with interest, at the end of five years to purchase a replacement tractor at the projected price of $28,050.

These calculations are based on interest paid on the sinking fund savings account being compounded annually. It is not uncommon for interest on savings accounts to be compounded quarterly or even more often. If interest on our sinking fund of the above example were compounded quarterly, the required annual deposit would be somewhat less. Interest would be compounded during a total of 20 periods (five years times four quarters per year) and the rate of interest per period would be five percent per year divided by four quarters per year or 1.25 percent. Table 8.6 does not contain entries for 1.25 percent but consulting collections of financial tables, such as the one by Gushee (1968), we obtain an Annual Deposit to Yield One value of 0.04432. The quarterly deposit required to yield funds sufficient to purchase the new tractor at the end of five years is $28,050 times 0.04432 or $1243.18.

FUTURE WORTH OF ANNUAL DEPOSITS

Questions answered by the sinking fund concept might be posed in a simple but equally relevant form: If I save $1000 per year in a savings account at five percent interest compounded annually, how much cash will I have accumulated at the end of five years? This question asks for the future worth of a flow of future annual deposits for which the general formula is

$$V_n = V_a \frac{[(1+i)^n - 1]}{i} = \$1000 \frac{[(1.05)^5 - 1]}{0.05},$$

where V_n is the future worth of equal annual amounts, V_a, deposited at the end of the year for n years in an account earning i rate of interest, compounded annually. The appropriate Future Worth of Annual Deposits value from Table 8.7 may be substituted for $[(1+i)^n - 1]/i$, giving

$$V_n = V_a(\text{Future Worth of Annual Deposits}) = \$1000(5.5256) = \$5525.60.$$

Thus the $1000 deposits I have made at the end of each of the 5 years have amounted to $5525.60 by the end of the fifth year. The Future Worth of Annual Deposits concept has its most important application in calculating the future worth of the savings a would-be investor makes toward the required down payment for a ranch, home, or other investment. Future worth calculations can be very discouraging to the hopeful young investor. A down payment of around 20 percent is usually required for the purchase of a home. With housing prices increasing as much as 10 percent annually in many areas, accumulation of a down payment is often a slow and painful process for the young family making small annual deposits to a savings account paying perhaps 5 or 6 percent interest compounded annually.

Table 8.7 Future Worth of Annual Deposits[a]

Years	3%	4%	$4\frac{1}{2}$%	5%	$5\frac{1}{2}$%	6%	$6\frac{1}{2}$%	7%	$7\frac{1}{2}$%	8%	9%
1	1.0000	1.0000	1.0000	1.0000	1.0000	1.0000	1.0000	1.0000	1.0000	1.0000	1.0000
2	2.0300	2.0400	2.0450	2.0500	2.0550	2.0600	2.0650	2.0700	2.0750	2.0800	2.0900
3	3.0909	3.1216	3.1370	3.1525	3.1680	3.1836	3.1992	3.2149	3.2306	3.2464	3.2781
4	4.1836	4.2464	4.2781	4.3101	4.3422	4.3746	4.4071	4.4399	4.4729	4.5061	4.5731
5	5.3091	5.4163	5.4707	5.5256	5.5810	5.6370	5.6936	5.7507	5.8083	5.8666	5.9847
6	6.4684	6.6329	6.7168	6.8019	6.8880	6.9753	7.0637	7.1532	7.2440	7.3359	7.5233
7	7.6624	7.8982	8.0191	8.1420	8.2668	8.3938	8.5228	8.6540	8.7873	8.9228	9.2004
8	8.8923	9.2142	9.3800	9.5491	9.7215	9.8974	10.0768	10.2598	10.4463	10.6366	11.0284
9	10.1591	10.5827	10.8021	11.0265	11.2562	11.4913	11.7318	11.9779	12.2298	12.4875	13.0210
10	11.4638	12.0061	12.2882	12.5778	12.8753	13.1807	13.4944	13.8164	14.1470	14.4865	15.1929
11	12.8077	13.4863	13.8411	14.2067	14.5834	14.9716	15.3715	15.7835	16.2081	16.6454	17.5602
12	14.1920	15.0258	15.4640	15.9171	16.3855	16.8699	17.3707	17.8884	18.4237	18.9771	20.1407
13	15.6177	16.6268	17.1599	17.7129	18.2867	18.8821	19.4998	20.1406	20.8055	21.4952	22.9533
14	17.0863	18.2919	18.9321	19.5986	20.2925	21.0150	21.7672	22.5504	23.3659	24.2149	26.0191
15	18.5989	20.0235	20.7840	21.5785	22.4086	23.2759	24.1821	25.1290	26.1183	27.1521	29.3609
16	20.1568	21.8245	22.7193	23.6574	24.6411	25.6725	26.7540	27.8880	29.0772	30.3242	33.0033
17	21.7615	23.6975	24.7417	26.8403	26.9964	28.2128	29.4930	30.8402	32.2580	33.7502	36.9737
18	23.4144	25.6454	26.8550	28.1323	29.4812	30.9056	32.4100	33.9990	35.6773	37.4502	41.3013
19	25.1168	27.6712	29.0635	30.5390	32.1026	33.7599	35.5167	37.3789	39.3531	41.4462	46.0184
20	26.8703	29.7780	31.3714	33.0659	34.8683	36.7855	38.8253	40.9954	43.3046	45.7619	51.1601
21	28.6764	31.9692	33.7831	35.7192	37.7860	39.9927	42.3489	44.8651	47.5525	50.4229	56.7645
22	30.5367	34.2479	36.3033	38.5052	40.8643	43.3922	46.1016	49.0057	52.1189	55.4567	62.8733
23	32.4528	36.6178	38.9370	41.4304	44.1118	46.9958	50.0982	53.4361	57.0278	60.8932	69.5319
24	34.4264	39.0826	41.6891	44.5019	47.5379	50.8155	54.3546	58.1766	62.3049	66.7647	76.7898
25	36.4592	41.6459	44.5652	47.7270	51.1525	54.8645	58.8876	63.2490	67.9778	73.1059	84.7008

Year											
26	38.5530	44.3117	47.5706	51.1134	54.9659	59.1563	63.7153	68.6764	74.0762	79.9544	93.3239
27	40.7096	47.0842	50.7113	54.6691	58.9891	63.7057	68.8568	74.4838	80.6319	87.3507	102.7231
28	42.9309	49.9675	53.9933	58.4025	63.2335	68.5281	74.3325	80.6976	87.6793	95.3388	112.9682
29	45.2188	52.9662	57.4230	62.3227	67.7113	73.6397	80.1641	87.3465	95.2552	103.9659	124.1353
30	47.5754	56.0849	61.0070	66.4388	72.4354	79.0581	86.3748	94.4607	103.3994	113.2832	136.3075
31	50.0026	59.3283	64.7523	70.7607	77.4194	84.8016	92.9892	102.0730	112.1543	123.3458	149.5752
32	52.5027	62.7014	68.6662	75.2988	82.6774	90.8897	100.0335	110.2181	121.5659	134.2135	164.0369
33	55.0778	66.2095	72.7562	80.0637	88.2247	97.3431	107.5357	118.9334	131.6833	145.9506	179.8003
34	57.7301	69.8579	77.0302	85.0669	94.0771	104.1837	115.5255	128.2587	142.5596	158.6266	196.9823
35	60.4620	73.6522	81.4966	90.3203	100.2513	111.4347	124.0346	138.2368	154.2516	172.3168	215.7107
36	63.2759	77.5983	86.1639	95.8363	106.7651	119.1208	133.0969	148.9134	166.8204	187.1021	236.1247
37	66.1742	81.7022	91.0413	101.6281	113.6372	127.2681	142.7482	160.3374	180.3320	203.0703	258.3759
38	69.1594	85.9703	96.1382	107.7095	120.8873	135.9042	153.0268	172.5610	194.8569	220.3159	282.6297
39	72.2342	90.4091	101.4644	114.0950	128.5361	145.0584	163.9736	185.6402	210.4711	238.9412	309.0664
40	75.4012	95.0255	107.0303	120.7997	136.6056	154.7619	175.6319	199.6351	227.2565	259.0565	337.8824
41	78.6632	99.8265	112.8466	127.8397	145.1189	165.0476	188.0479	214.6095	245.3007	280.7810	369.2918
42	82.0231	104.8195	118.9247	135.2317	154.1004	175.9505	201.2711	230.6322	264.6983	304.2435	403.5281
43	85.4838	110.0123	125.2764	142.9933	163.5759	187.5075	215.3537	247.7764	285.5506	329.5830	440.8456
44	89.0484	115.4128	131.9138	151.1430	173.5726	199.7580	230.3517	266.1208	307.9669	356.9496	481.5217
45	92.7198	121.0293	138.8499	159.7001	184.1191	212.7435	246.3245	285.7493	332.0645	386.5056	525.8587
46	96.5014	126.8705	146.0982	168.6851	195.2457	226.5081	263.3356	306.7517	357.9693	418.4260	574.1860
47	100.3965	132.9453	153.6726	178.1194	206.9842	241.0986	281.4525	329.2243	385.8170	452.9001	626.8627
48	104.4083	139.2632	161.5879	188.0253	219.3683	256.5645	300.7469	353.2700	415.7533	490.1321	684.2804
49	108.5406	145.8337	169.8593	198.4266	232.4336	272.9584	321.2954	378.9989	447.9348	530.3427	746.8656
50	112.7968	152.6670	178.5030	209.3479	246.2174	290.3359	343.1796	406.5289	482.5299	573.7701	815.0835

[a]Future value of $1 deposited annually at end of year.

CREDIT DECISIONS—WHERE TO BORROW FUNDS

There is considerable variability among lending institutions in both interest rates charged and loan repayment schedules. "Comparison shopping" for borrowed funds should always be based on a single universal parameter, the simple annual interest rate, commonly called the *annual percentage rate* (APR). Truth-in-lending legislation was designed to enable borrowers to make accurate comparisons between sources of funding. Credit card and other revolving credit invoices, for example, state that interest charges will be "18 percent interest per annum." However, disclosure of APR is often made in fine print on the loan agreement copy that the borrower receives after the loan has been approved. While conforming to truth-in-lending laws, such disclosure is of little help in comparison shopping. Fortunately, a simple formula allows lending contract terms to be compared on the basis of simple annual interest rate prior to APR disclosure required by law. The general formula for calculating the simple annual interest rate, i, is

$$i = \frac{2NI}{B(n+1)},$$

where N is the number of payments made per year, I is the total interest paid (in dollars), B is the total amount borrowed (the amount actually available for use by the borrower), and n is the total number of payments made during the entire life of the loan.

Suppose a rancher needs to replace his old pickup truck. The new pickup is priced at $10,500 and the old one has a trade-in value of $1500. Both his bank and his credit union have agreed to lend the needed $9000 at "$7 interest per year per $100 borrowed—or about seven percent" with monthly repayment over a $2\frac{1}{2}$-year period. The two loans are not identical, however. The credit union loan terms are based on the *add-on method*. Monthly payments are calculated by adding interest charges ($7 per $100 per year times $9000 times $2\frac{1}{2}$ years = $1575) to the $9000 principal and dividing by 30 months, resulting in monthly payments of $352.50. The face value of the credit union note is the total of principal and interest, $10,575.

The bank establishes loan terms according to the *discount method*. Interest charges are calculated at $1575, just as they were by the credit union. But the bank provides for all interest due to be paid at the time the loan is initiated and $1575 is subtracted from the $9000 face value of the note leaving $7425 as the principal amount actually available for purchase of the pickup.[11] Monthly loan payments are $300.00, the result of dividing the $9000 face value of the note by 30 months.

Despite the fact that $9000 was requested from each lending institution and the fact that both lenders have quoted $7 per $100 per year as the interest charges and agreed to make 30 month loans, no meaningful comparison of loan terms can be made until both loans are reduced to a single parameter, simple annual interest.[12] Substituting terms for the credit union add-on loan into the general formula:

$$i = \frac{2(12)1575}{9000(31)} = 0.1355.$$

Similarly, for the bank discount loan:

$$i = \frac{2(12)1575}{7425(31)} = 0.1642.$$

Thus the simple annual interest rate charged on the credit union loan is actually almost three percent less than that for the bank loan. While not as accurate as the APRs generated by modern financial calculators (interest rates of 12.88 percent for the credit union loan and 15.47 percent for the bank loan were calculated with a Hewlett-Packard 22), this formula is fast and easy to work with and provides sufficient accuracy for most decisions involving choice of loan terms.

LITERATURE CITED

Gushee, C. H. 1968. *Financial Compound Interest and Annuity Tables.* Financial Publishing Co., Boston, MA. 884 pp.

Nielsen, D. B. 1967. *Economics of Range Improvements—A Ranchers Guide to Economic Decision Making.* Utah Agricultural Experiment Station Bulletin No. 466. 48 pp.

Nielsen, D. B. and J. P. Workman, 1971. *The Importance of Renewable Grazing Resources on Federal Lands in the 11 Western States.* Utah Agricultural Experiment Station Circular No. 155. 44 pp.

Workman, J. P. and J. F. Hooper. 1975. Impact of certain federal income tax provisions on rangeland development and rangeland prices. *Abstracts of Papers of 28th Annual Meeting, Society for Range Management. Mexico City, Mexico, February 13,* 1975. Society for Range Management, P. 38.

TEXT NOTES

1 I have occasionally encountered students who claimed a preference for the future dollar because they would be forced to save money that would otherwise be wasted on unimportant purchases.

2 Inflation effects can be avoided in discounting calculations by expressing projected future costs and benefits in real (inflation-free) prices. Risk can be handled simply by increasing project implementation and maintenance costs to cover possible failures.

3 In the present example interest is compounded annually so each period is one year in length. If interest were compounded semiannually, each period would be six months.

4 Provided, of course, that interest and principal payments are not due until after you sell the land.

5 Again, from the standpoint of cash flow this is a feasible investment only if loan payments are not due until after you sell the land.

6 The nominal bank borrowing rate is much higher than 4 percent, perhaps 10

percent. The nominal rate includes the real opportunity cost rate of 4 percent plus 6 percent expected inflation. However, since our analysis is based on noninflated (real) current returns of $7.50 per AUM, the real borrowing rate is the appropriate rate for our discounting calculations. This point is explored further in Chapter 9.

7 The range manager is often understandably disturbed by the conclusions provided by the "dismal science" of economics. Here is a reseeding project that the range manager knows will double carrying capacity and the economist's analysis tells him he cannot afford it! However, the economist's recommendations are not always this dreary. Suppose the above rancher also had a depleted 100-acre tract with a current carrying capacity of 4 acres per AUM and that reseeding could also be expected to double forage production on this site. While the cost of reseeding this more productive tract would remain at $18 per acre, the expected annual returns would be considerably higher. Substituting the appropriate numbers into the given formula, the present value of 25 additional AUM annually for 20 years is $2548.13, yielding a $748.13 net profit. These calculations illustrate the importance of concentrating range improvement funds on the more productive and responsive sites.

8 Net income to land and improvements is net ranch income described in Chapter 2 minus both operator and family labor and interest on investment in livestock and machinery. It is the income attributable solely to ownership of real estate.

9 The rationale underlying this calculation is that as long as the investor receives sufficient income to pay interest on borrowed funds and to pay opportunity cost on his down payment, he would be indifferent between investing in the ranch and in any other opportunity. Since the principal payment on the loan is a payment from the investor to himself (in the form of increased equity), there is no need to include the principal payment in the calculation.

10 Again it should be noted that these calculations are based on payment of loan principal and interest at the end of the 10-year period after the ranch has been sold. Since interest and principal payments are normally made annually, the maximum price would be somewhat less than $194,896.

11 If not anticipated, this may be disturbing news for the borrower, since he must now pay $1575 down on the pickup purchase in addition to his $1500 trade-in. Even if the borrower is aware of how the bank's discount loan arrangement works, he must calculate how much the face value of the note must be in order to avoid the additional down payment. Since a total of $9000 usable principal is required and interest is calculated at seven percent per year for $2\frac{1}{2}$ years, the necessary note face value can be found by solving the following equation:

$$X - 0.07X(2.5) = \$9000 \quad \text{or} \quad X = \$10{,}909.$$

12 The reader may wonder why the percent per year interest rates quoted by lending institutions are not equivalent to simple annual interest rates. Referring to the loan terms offered by the credit union, the total dollar interest charge of $1575 is based on the original amount borrowed rather than the outstanding balance remaining at the beginning of a particular month. Although the borrower does owe $9000 at the beginning of the first month, by the beginning of month 30, the principal amount outstanding has been reduced to less than $352.50. In multiplying $7 per $100 per year by $2\frac{1}{2}$ years, the credit union actually charges interest on principal long after it has been repaid. This causes the simple annual interest rate to greatly exceed the stated rate.

Chapter 9

Economic Analysis of Private Range Improvements

THE ANALYSIS — A GENERAL OVERVIEW

Almost all range improvement projects have expected useful lives of more than one year. In its simplest form, economic analysis of these projects consists of comparing per acre treatment costs with the present value of the flow of annual per acre net returns. The hypothetical data in Figure 9.1 for a prescribed burning project demonstrate an idealized schedule of costs and returns encountered with range improvements. The burning project requires a relatively large initial investment (I) of $4 per acre. Real (noninflated) net annual returns (R) are a much smaller $0.50 per acre, but they are received over a 15-year life (n). To test the economic feasibility of this project we must know the real[1] interest rate on borrowed funds and the real opportunity cost for owned capital. If funds can be borrowed at a real rate of four percent and if no alternative investment opportunity promises to yield more than a four percent real return, we can now apply the appropriate Present Worth of One Per Period value (Table 8.5):

$V_0 = R$ (Present Worth of One Per Period Value for 15 years and four percent) or

$V_0 = \$.50(11.118) = \$5.56$.

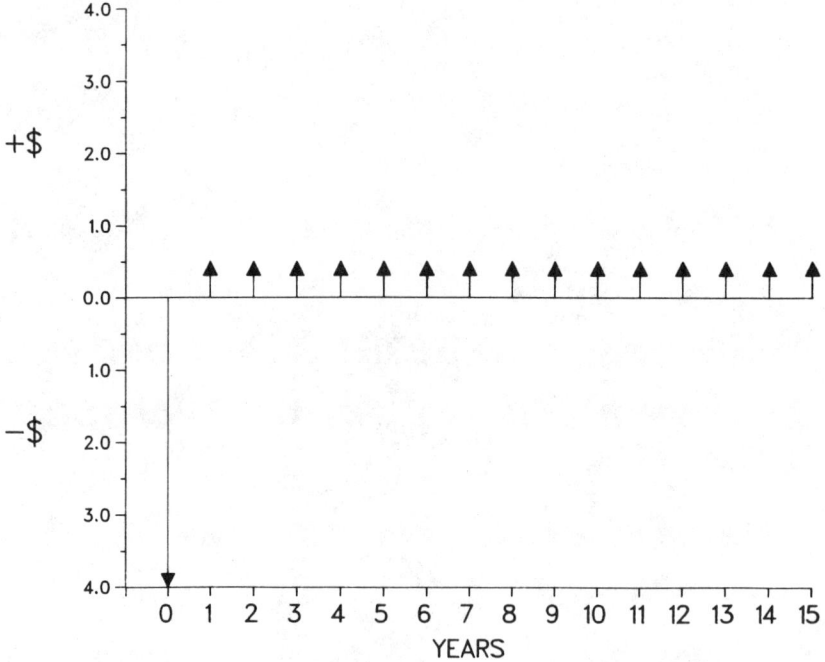

Figure 9.1 Required investment and net annual returns for a hypothetical prescribed burning project with a 15-year life.

We now have the future annual net returns expressed in current dollars ($5.56) and this amount can be compared directly with the initial investment ($4.00) to determine whether or not the project is economically feasible.

Present Net Worth

One such comparison is present net worth (PNW) calculated as $5.56 − $4.00 = $1.56 per acre. This positive PNW may be interpreted to mean that if funds for the project can be borrowed at four percent real interest and if the best alternative use of borrowed and owned funds does not promise a real return greater than four percent over its 15-year life, the project will yield a profit of $1.56 per acre treated. By the PNW criterion the improvement is economically feasible.

Benefit–Cost Ratio

Another comparison widely used by land management agencies of the federal government is the benefit–cost ratio (B/C). Using data from Figure 9.1, this ratio is calculated by dividing the present value of the annual benefits (B), $5.56, by the present value of initial investment and annual project maintenance costs (C), $4.00, to form ($B/C$) = 1.39.[2] The B/C ratio may be interpreted to mean that if range improvement funds can be borrowed at four percent real interest and if no alternative use of these funds offers a real return greater than four percent, the project is economically feasible since it will return the present value equivalent of $1.39 for each $1.00 invested.

Internal Rate of Return

A third comparison commonly employed in decisions involving improvement of private rangeland is the internal rate of return (IRR) defined in Chapter 8 as "the interest rate that discounts or forces a future stream of net returns to just equal the investment required to produce the flow of returns" (Nielsen, 1967). IRR is also based on the formula for discounting future flows of annual income.[3] Calculation of IRR for the example of Figure 9.1 involves setting the initial investment (I) equal to the formula for present value of annual net returns:

$$I = R \frac{[1-(1+i)^{-n}]}{i}.$$

Substituting values from Figure 9.1,

$$\$4 = \$0.50 \frac{[1-(1+i)^{-15}]}{i} \quad \text{or}$$

$$\frac{\$4}{\$0.50} = 8 = \frac{[1-(1+i)^{-15}]}{i}.$$

The right hand portion of the latter equation corresponds to the Present Worth of One Per Period value for 15 years (Table 8.5) for the unknown i (IRR) generated by the prescribed burning project. Referring to Table 8.5, our hypothetical project yields an IRR of just over nine percent.[4] Interpretation is as follows: If funds can be borrowed at less than nine percent real interest and if no alternative investment promises a real return greater than nine percent, the project is economically feasible.

For the borrowing and opportunity cost rates previously specified, conclusions reached by all three investment criteria are that the hypothetical prescribed burn is an economically feasible project.[5] We now turn to a detailed discussion of procedures used to estimate the data required for economic analyses of range improvement projects.

INFORMATION REQUIRED FOR ANALYSIS

Information required for economic analysis of range improvements on privately owned rangeland include: (1) quantity of expected benefits on specific sites selected for treatment; (2) value of expected benefits; (3) expected productive life of project; (4) expected costs, including both original investment and maintenance or operating costs induced by the project; and (5) the interest rate paid for borrowed capital and the rate of return on the best alternative investment opportunity.

Expected Project Benefits

The primary objective of most range improvement projects is to produce more usable forage resulting in (1) increased yearlong carrying capacity making herd expansion possible and (2) decreased feed costs per animal unit through substitution of range forage for more expensive home grown or purchased feeds. Herd expansion increases gross ranch income but even more important, the reduction in feed costs per animal unit increases net ranch income. These simultaneous effects are summarized in the "unit cost surface" of Figure 9.2. This three-dimensional graph shows the production costs per pound of beef (height of the box) as a function of carrying capacity in acres per AUM and ranch size in number of brood cows run (the two sides of the box). If a rancher were operating at *A* (200 head of brood cows and an average range carrying capacity of 8 acres per AUM), he could decrease per pound beef production costs in three ways. First, he could expand herd size from 200 to 300 cows by purchasing or leasing more of the "8-acre" land that he is currently running on (a move from *A* to *B* on the graph). This might be called his "bigger is better" option, which takes advantage of the economies of size reported for cattle ranching operations (Martin and Goss, 1963; Workman and Hooper, 1971).

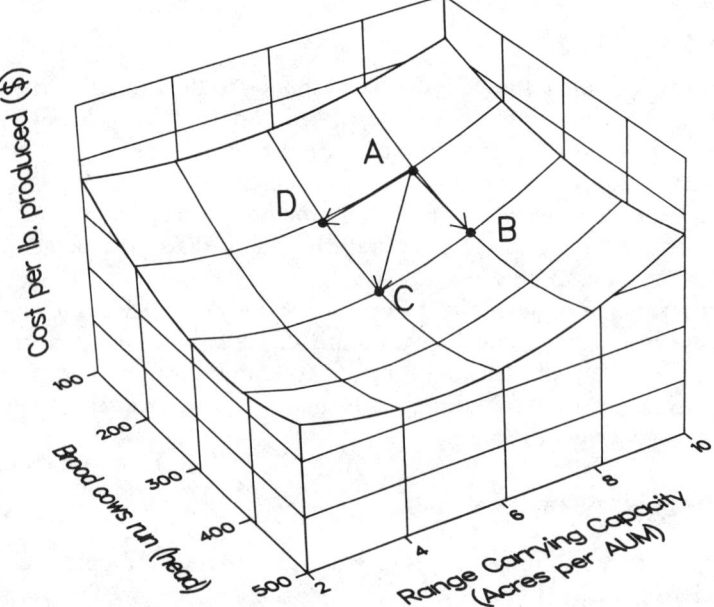

Figure 9.2 Hypothetical effects of carrying capacity and ranch size on unit production costs (after Martin and Goss, 1963).

Second, he might move his existing 200 head operation to a more productive land base by selling his 8-acre land and purchasing 6-acre land or by improving his existing range to a carrying capacity of 6 acres per AUM (a move from A to D). This move could be called his "move or improve" option. Third, improvement of current rangeland may well make possible a move from A to D since increased carrying capacity not only reduces unit production costs but in some cases alleviates seasonal forage constraints and allows an expansion in herd size. Thus a move from A to C would provide the largest increase in net ranch income since it increases profit per pound sold *and* allows more pounds to be sold.

One of the easiest and most effective means of estimating the increased forage production expected to result from range improvements involves the use of Yield and Vegetation Composition guides published by the USDA Soil Conservation Service (SCS) as part of their range analysis work on private rangeland (Renner and Allred, 1962). A *range site* is a particular kind of rangeland (usually distinguished from other range sites on the basis of elevation, soil texture and depth, average annual precipitation, and topography) capable of producing a specific potential amount and kind of forage. SCS Yield and Vegetation Composition guides form the basis of a wealth of range management information concerning ranches owned or operated by SCS "cooperators." Individual ranchers cooperating in the SCS program receive a 2-inch = 1-mile aerial photograph of their ranch upon which fences, buildings, roads, cropland soils, range sites, and range condition,[6] by site and by pasture, have been delineated. Determination of the number of acres in each range condition class for each range site through the use of a planimeter or dot grid, combined with Yield and Vegetation Composition guides, allows average forage production of the ranch to be estimated. For example, suppose we are planning a prescribed burn to improve 417 acres of rangeland delineated on the ranch aerial photo as the Upland Loam range site, presently in fair condition. Referring to Figure 9.3, we find that this particular site in fair condition is expected to produce an average of 1527 pounds of air-dry herbage per acre in favorable years and an average of 842 pounds in unfavorable years. If favorable years can be expected to occur in about the same frequency as unfavorable years, annual herbage production averages about 1185 pounds per acre. Since not all vegetation is forage, herbage must now be converted to useable forage. According to Figure 9.3, the Upland Loam site in fair condition is composed of only 58 percent forage species, so estimated annual air-dry forage production is about 687 pounds per acre. Because of sustained yield considerations, not all of this forage can be grazed. Applying the common range management rule of "take half–leave half," only 344 pounds per acre of useable forage are produced annually prior to the prescribed burn.[7]

Now suppose that, with the help of the SCS range conservationist, we estimate that the prescribed burn will reduce woody vegetation and encourage expansion of perennial native grasses, allowing range condition of this site to improve from fair to good condition. Through calculations like those in the

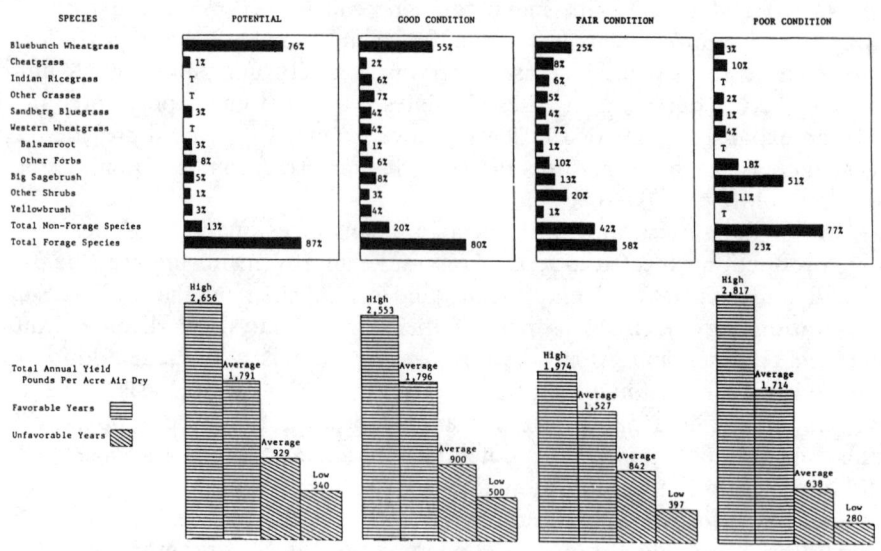

Figure 9.3 Yield and Vegetation Composition.

preceding discussion, we determine that the expected usable annual air-dry forage production after improvement is 539 pounds per acre, an increase of 195 pounds.

Before this forage production increase can be valued in terms of either its direct market value or its potential contribution to livestock production, it must be converted to animal unit month (AUM) feed equivalents. An AUM may be defined as the amount of feed required per month by one animal unit (usually considered to be a 1000-pound cow or her equivalent) for maintenance and normal growth. According to the National Research Council (1970), a lactating 1000-pound cow requires an average of about 22 pounds of dry matter daily (or about 660-pounds air-dry usable forage per month).[8] Dividing the 195 pounds of additional forage per acre due to the prescribed burn by the 660 pounds required per AUM yields an estimated 0.30 additional AUM per acre or a total of 125 AUM for the 417-acre tract of Upland Loam.

Value of Expected Benefits

Before being compared with project costs, the benefits expected from a range improvement must be converted to increased dollar returns. Numerous analytical techniques have been employed by researchers to value additional AUM of forage. These techniques fall into two general categories: (1) a simple approach seeking to value added AUM in terms of their direct market value and (2) a more complicated technique that attempts to estimate the value of increased livestock production made possible by additional AUM.

Forage Value Based on Private Lease Rate Straightforward and much easier to use, the first technique is sufficiently accurate in some situations. Nielsen and Hinckley (1975), for example, valued increased AUM from sagebrush control at the current average private range lease rate. If a rancher is currently leasing private range and can avoid paying future lease fees through improvement of his own rangeland, this simple evaluation procedure is quite valid. Even the rancher who does not normally lease range from other owners usually has the opportunity to lease forage produced by range improvements to neighboring stockmen. Thus in our prescribed burning example the 0.30 additional AUM per acre annually might have been valued in 1980 at the average private lease rate of $8.00 per AUM (USDA, 1980) resulting in an added return of $2.40 per acre per year.[9]

A more complicated approach to calculating the average value of an added AUM has been used by Kearl (1975). All nonforage costs were subtracted from annual gross ranch income and the result divided by the total number of AUM of forage used by the operation annually. Kearl preferred this method to that of applying the average private lease rate, since the ranches he studied normally did not graze leased private rangeland.

An alternative to the two average value methods is based on the premise that due to crucial shortages of forage during certain seasons of the year, both the average lease value of an AUM and the amount of net income attributable to the average AUM have little meaning to some ranching operations (McCormick and Workman, 1975). Instead additional AUM that become available during a particular season, such as early spring, are valued in terms of avoided costs of purchased alternative feeds, such as hay. Again, if purchased hay is required in the current ranch operation and if hay purchases avoided are the only benefits resulting from a range improvement, this approach offers an accurate measure of forage value. However, in many ranching situations seasonal forage imbalance is sufficiently serious that there are forage surpluses in some seasons and forage shortages in others. Thus range improvements may make possible the use of a previously wasted surplus forage resource as well as provide cost savings through decreased requirements for purchased feeds. Use of previously surplus forage may allow expansion of the base herd and increased production of salable livestock. Similarly, alleviation of seasonal forage shortages may sufficiently improve the nutritional status of livestock to allow increased conception and birth rates and improved weaning weights. For these two reasons, the more complicated technique of valuing the contribution of additional forage to livestock production is often the only means of accurately accounting for total range improvement benefits. Cook et al. (1980) have used linear programming analysis to more precisely estimate the value of public land forage.

Forage Value Based on Livestock Production

Seasonal Forage Balance Estimation of forage value due to increased livestock production from range improvements requires an assessment of changes in livestock forage balance. The assessment begins with the construction of two charts. The first, called a forage or feed balance chart, is a tabular representation of monthly or seasonal availability of all sources and amounts of feed on the ranch; the second, the stock count chart, portrays the monthly feed requirements by class of livestock. Prior to construction of the necessary charts, all quantities of available and required feed must be converted to AUM. Suppose the available feeds for the 300 head cattle ranch of Chapter 2 are as shown in Table 9.1.[10] Barley commonly contains about 75 percent total digestible nutrients (TDN) and mixed meadow hay contains about 50 percent.[11] Combining this information with the rule of thumb that 400 pounds of TDN are required per AUM (American Institute of Real Estate Appraisers, 1972) and remembering that the average weight of barley is 48 pounds per bushel, the contribution of the barley is calculated to be

$$\frac{2000 \text{ bushels} \times 48 \text{ pounds} \times 0.75 \text{ TDN}}{400 \text{ pounds TDN/AUM}} = 180 \text{ AUM}.$$

Similarly, the meadow hay represents

$$\frac{784 \text{ tons} \times 2000 \text{ pounds} \times 0.5 \text{ TDN}}{400 \text{ pounds TDN/AUM}} = 1960 \text{ AUM}.$$

Now that all available feeds are expressed in terms of AUM we can begin to allocate the feed supply among months in our trial feed balance chart (Table 9.2). This will be a preliminary allocation, since monthly feed requirements in the form of a stock count chart have not yet been made. We begin with the inflexible feed sources from Table 9.1. Forage from native range, seeded pasture, and barley and meadow aftermath (the forage remaining in fields and meadows after grain and hay has been harvested) are relatively inflexible feeds as compared with barley and hay. Due to timing of range readiness and susceptibility to injury during the initial portion of the grazing season, the 1700

Table 9.1 Feeds Available on a Hypothetical 300-cow Utah Cattle Ranch

Feed source	Quantity	Unit
Native range	1700	AUM
Seeded pasture	280	AUM
Barley	2000	bushels
Meadow hay	784	tons
Aftermath	554	AUM

Source: Workman and MacPherson (1973).

Table 9.2 Trial Feed Balance Chart for a Hypothetical 300-Head Utah Cattle Ranch (AUM)

Month	Source of feed					Total available
	Range	Seeded pasture	Barley	Meadow hay	Aftermath	
Jan.						
Feb.						
Mar.						
Apr.						
May	212	140				
June	213	140				
July	425					
Aug.	425					
Sept.	425					
Oct.					278	
Nov.					277	
Dec.						
Total	1700	280	180	1960	555	4675

Source: Workman and MacPherson (1973).

AUM of native forage cannot be evenly distributed over the five month grazing season. Instead only about half as much native forage can be harvested monthly during May and June as during the other three months. Based on this utilization constraint, range forage is allocated in Table 9.2 subject to any changes made necessary by the monthly distribution of feed requirement to be calculated in Table 9.3. Seeded pasture represents a purposely inflexible feed supply, since a cool season wheatgrass pasture has been established on the subject ranch specifically to help alleviate the shortage of native spring range. Availability of seeded pasture is evenly divided between the months of May and June. Inflexibility of crop and meadow aftermath is simply due to the fact that this forage must be used after crop harvest and before being covered with snow. Prior to development of the stock count chart, aftermath is initially evenly divided between the months of October and November. Except for the early spring season (to be discussed), barley and hay can be freely allocated among months and the trial feed balance chart is as complete as we can make it prior to construction of the stock count chart.

The first step in formulating the stock count chart is to convert all the various kinds and ages of livestock to a common measure, the animal unit (AU). As mentioned, an AU is usually defined as a 1000-pound cow or her equivalent.[12] Cattle "equivalents" have often been calculated as 0.1 AU for each 100 pounds of live cattle weight (American Institute of Real Estate Appraisers, 1972). In the case of sheep equivalents, the U.S. Forest Service has long used a conversion ratio of five sheep equal one cow or 1 AU. Less efficient than ruminant livestock, AU equivalents for horses are often calculated as 1.5 AU per 1000 pounds. Currently, the most widely recommended procedure is to

Table 9.3 Stock Count Chart for a Hypothetical 300-Head Utah Cattle Ranch

Month	Cows (1.0 AU)		Yearling Heifers (0.74 AU)		Calves (0.42 AU)		Bulls (1.29 AU)		Total required AUM
	Head	AUM	Head	AUM	Head	AUM	Head	AUM	
Jan.	300	300			45	18.9	15	19.4	338
Feb.	300	300			45	18.9	15	19.4	338
Mar.	300	300			45	18.9	15	19.4	338
Apr.	300	300	45	33.3	born		15	19.4	353
May	300	300	45	33.3			15	19.4	353
June	300	300	45	33.3			15	19.4	353
July	300	300	45	33.3			15	19.4	353
Aug.	300	300	45	33.3	240	100.8	15	19.4	454
Sept.	300	300	45	33.3	240	100.8	15	19.4	454
Oct.	300	300	45	33.3	240	100.8	15	19.4	454
Nov.	300	300			45	18.9	15	19.4	338
Dec.	300	300			45	18.9	15	19.4	338
Total									4464

Source: Workman and MacPherson (1973).

form a metabolic requirement ratio $(W)^{0.75}/(1000 \text{ pounds})^{0.75}$, where W is the weight (in pounds) of the animal in question and a 1000-pound cow is defined as the basic AU. Thus a 150-pound ewe represents $(150)^{0.75}/(1000)^{0.75} = 0.25$ AU, and four rather than five sheep are equivalent to one cow.

From Chapter 2, we have the information necessary for the construction of our stock count chart (Table 9.3). We begin with the cow column, knowing that 300 breeding cows are run and that 15 percent (45 head) are culled annually in November just after pregnancy evaluation. Also in November, 45 bred long yearling replacement heifers enter the herd, maintaining the breeding herd at 300 head during each month of the year. Turning now to the calf column, 240 calves are born during March, April, and May. On August 1, four months after their average birth date, the calves are assumed to begin using forage, independent from that taken by their mothers. At this time the calves weigh about 250 pounds and represent $(250)^{0.75}/(1000)^{0.75} = 0.35$ AU each. At weaning time on November 1, the calves average 390 pounds or 0.49 AU. For the three months August 1–November 1, each calf requires an average of 0.42 AUM of forage per month and the total monthly AUM requirement for calves is based on this figure. After weaning, 195 calves are sold and 45 heifer calves remain in the calf column until they become yearlings on April 1 of the following year. At this time they weigh about 550 pounds each. By November 1 they have grown to about 800 pounds and they are entered into the cow column to replace cows culled from the herd. Thus, on the average during this seven month period, they require 0.74 AUM per head per month. Finally the bulls are entered into the chart. From Chapter 2, 15 bulls are run or 1 bull for each 23 cows or yearling replacement heifers present in the herd during the breeding season. Since the 5 bulls culled from the herd in November are immediately replaced with young purchased stock, 15 bulls, weighing an average of 1400 pounds and requiring 1.29 AUM of feed monthly, are present during each month. Total monthly AUM requirements are now determined by multiplying the number of head in each animal class by the appropriate AU requirement and summing across each row. For January, 300 cows × 1.0 AU + 45 calves × 0.42 AU + 15 bulls × 1.29 AU = 338.3 AU present for one month or 338.3 AUM.

We are now ready to combine the information in the stock count chart (Table 9.3) with that in the trial forage balance chart (Table 9.2) in order to formulate a final forage balance chart. This final chart (Table 9.4) will reveal the degree of balance between feed available and feed required and will let us determine whether the yearlong carrying capacity of the ranch in question really is 300 cows. We begin the construction of Table 9.4 by filling in the Total Required AUM column of Table 9.3 along with the bottom row from Table 9.2, which shows the number of AUM supplied by each feed source. Feeds from all sources, subject to Table 9.2 constraints on availability of the inflexible feeds (range, pasture, and aftermath), are now allocated to fit the monthly feed needs shown in the Total Required column. During May and June, the allowable use of native range plus the seeded pasture is just sufficient to meet feed needs. Forage requirements and supplies can also be made to balance during July, August, and September by reducing the July native range allocation from 425 to 355 and dividing the 70 AUM released evenly between August and September to bring the available AUM for each of these months to 460.

Table 9.4   Final Feed Balance Chart for a Hypothetical 300-Head Utah Cattle Ranch (AUM)

Month	Source of feed					Total required	Total available
	Range	Seeded pasture	Barley	Meadow hay	Aftermath		
Jan.			25	341		338	366
Feb.			25	341		338	366
Mar.			25	341		338	366
Apr.			25	356		353	381
May	212	140				353	352
June	213	140				353	353
July	355					353	355
Aug.	460					454	460
Sept.	460					454	460
Oct.			30		454	454	484
Nov.			25	240	101	338	366
Dec.			25	341		338	366
Total	1700	280	180	1960	555	4464	4675

Source: Workman and MacPherson (1973).

The next adjustment to the trial allocation of Table 9.2 is for October. For the purpose of "creep feeding" calves to be marketed November 1, 30 AUM of barley are allocated for the month of October leaving 25 AUM per month for November–April for "growing out" replacement heifers. The remainder of the October feed requirement is met by reallocating 176 AUM of November aftermath from Table 9.2 to October, bringing the total available AUM to 484 and leaving 101 AUM of aftermath for use during November. The remainder of the November feed requirement is met through the allocation of 240 AUM of hay, bringing the total monthly AUM available for November–March to 366.[13] Based on our completed forage balance chart, we can conclude that feed availability does match feed requirement and that the ranch in question is capable of supporting a 300 head breeding herd yearlong.

Examination of Table 9.4 also reveals that total yearlong carrying capacity is constrained by the limited forage resources available from May through September. During the other seven months of the year feed supplies are more than adequate. Also, even if shortages did exist during the period October–April, the deficit could be alleviated with purchased hay or grain. Purchased feeds do not offer a workable solution to shortages during the growing season, especially the early portion. Not only do cattle not relish hay and concentrates when green forage is available, but confinement feeding during May often leads to wet, muddy conditions giving rise to calf scours and other animal diseases. If these confinement problems are avoided by feeding hay and concentrates on the range, the natural preference of cattle for green forage from growing plants may lead to ingestion of poisonous plants since these species are usually the first plants to initiate spring growth. Thus, for the ranch in question, the forage supply available during the "limiting months" of May and June sets the maximum yearlong breeding herd carrying capacity.

Once the limiting months have been identified and the feed supply available during each month is known, yearlong breeding herd carrying capacity can be accurately calculated. Equally important when, as in this example, a figure for breeding herd size is already available, monthly feed supply information and knowledge of the limiting months allows maximum herd size carrying capacity to be easily verified. Although several rules of thumb approaches[14] have long been used by range managers and ranch appraisers to calculate or verify breeding herd carrying capacity, the algebraic method (Workman and MacPherson, 1973) appears to be much more accurate. This method first identifies the months that impose the most severe limits on yearlong breeding herd carrying capacity (the months during which available feed supplies are the smallest and that are composed of inflexible feeds). Next the feed required by livestock during the most limiting month is calculated in terms of X, the unknown maximum breeding herd that can be supported for that month. From Chapter 2 and Table 9.3, we know that during the limiting months of May and June the cattle herd consists of cows requiring 1 AUM per head, yearling replacement heifers (0.15 heifers per cow) requiring 0.74 AUM

ECONOMIC ANALYSIS OF PRIVATE RANGE IMPROVEMENTS

per head, and bulls (0.05 bulls per cow) requiring 1.29 AUM per head. The May feed requirement is

$$1.00 \text{ cow } (1.00 \text{ AUM}) X + 0.15 \text{ heifer } (0.74 \text{ AUM}) X \\ + 0.05 \text{ bull } (1.29 \text{ AUM}) X = 1.1755 X \text{ (AUM)}.$$

During May, then, each breeding cow and her necessary herd complement require 1.1755 AUM of feed. Feed requirement is now set equal to the feed available during May,

$$1.1755 X \text{ (AUM)} = 353 \text{ AUM},$$

and solving for X we obtain 300, the maximum number of cows that the ranch will support during May. Since May and June are the limiting months constraining yearlong cow herd carrying capacity, we know that feed supplies are adequate for 300 head of cows and their herd complement during the entire year.

We are now ready to address the first of two important questions concerning the economic evaluation of our prescribed burn: How many additional brood cows, along with the necessary herd complement of bulls, calves, and replacement heifers, can be supported yearlong due to the addition of 125 AUM of range forage produced through the proposed range improvement? This question is easily answered by revising the above forage balance and stock count charts. Since the only portions of the forage balance chart that will be changed by the increased range forage are the AUM available monthly for the period May–September, this information is easily combined with a

Table 9.5 Stock Count Chart for a Hypothetical Utah Cattle Ranch[a]

	Animal class								Total required AUM	Total available AUM
	Cows (1.0 AU)		Yearling heifers (0.74 AU)		Calves (0.42 AU)		Bulls (1.29 AU)			
Month	Head	AUM	Head	AUM	Head	AUM	Head	AUM		
Jan.	320	320			48	20.2	16	20.6	361	366
Feb.	320	320			48	20.2	16	20.6	361	366
Mar.	320	320			48	20.2	16	20.6	361	366
Apr.	320	320	48	35.5	born		16	20.6	376	381
May	320	320	48	35.5			16	20.6	376	377
June	320	320	48	35.5			16	20.6	376	378
July	320	320	48	35.5			16	20.6	376	380
Aug.	320	320	48	35.5	256	107.5	16	20.6	484	485
Sept.	320	320	48	35.5	256	107.5	16	20.6	484	485
Oct.	320	320	48	35.5	256	107.5	16	20.6	484	484
Nov.	320	320			48	20.2	16	20.6	361	366
Dec.	320	320			48	20.2	16	20.6	361	366
Total									4761	4800

[a]Revised to include 20 additional cows made possible by 125 additional AUM from range improvement.

revised stock count chart (Table 9.5). We first fill in the Total Available AUM column. The 125 additional AUM are spread evenly through the five month grazing season, increasing available AUM for each month by 25. All other figures for monthly feed available remain as in Table 9.4. Employing the algebraic method, we set the forage requirement per breeding cow during May ($1.755X$) equal to the feed available in May (377 AUM). Solving for X, our revised maximum breeding herd carrying capacity is 320.71, conservatively rounded to 320 cows or an increase of 20 head. Based on a revised breeding herd of 320 head, Table 9.5 can now be completed and Total Required AUM can be compared with Total Available AUM. This comparison reveals that the feed requirements of the expanded breeding herd are met by feed supplies available after improvement. Due to surplus fall and winter feed resources prior to improvement, the 125 AUM forage increase during the grazing season has increased total yearlong breeding herd carrying capacity by 20 cows.

The second important question to be answered now is: What will be the resulting increase in net ranch income brought about by the increased herd size made possible by the prescribed burning project?[15]

Partial Budgeting To convert the expected increase in breeding herd size to an increase in annual ranch income we must now refer back to Chapter 2 to obtain the necessary production and income information for our subject ranch. Referring to Table 2.2 we obtain the appropriate sale weights and prices for the additional cull cows, cull bulls, and weaner calves produced by the 20 additional breeding cows. The resulting additional gross ranch income is shown in Table 9.6. The $4132 addition to gross ranch income is only one of several revisions to the ranch income statements of Chapter 2 (Tables 2.3 and 2.4) that must be made in order to calculate the increase in net ranch income that will ultimately determine the economic feasibility of the prescribed burn.

Table 9.6 Hypothetical Added Annual Production and Gross Ranch Income from 20 Additional Cows[a]

Number and category	Sale weight (lb)	Sale price ($/lb)	Total sales ($)
3 cull cows	1000	0.35	1050.00
$\frac{1}{3}$ cull bull[b]	1500	0.40	200.00
5 heifer calves[c]	380	0.54	1026.00
8 steer calves	400	0.58	1856.00
Total added cash returns			$4132.00

[a]Made possible by 125 additional AUM from range improvement, 1980.

[b]The 20 additional cows require 1 additional bull and bulls are replaced every 3 years.

[c]Calf crop remains at 80 percent resulting in 16 additional calves from the 20 additional cows. Cow replacement rate is 15 percent, requiring that 3 of the 8 additional heifer calves be retained as replacements.

Expected changes in annual operating costs must also be accounted for and these appear in Table 9.7. This procedure of modifying the ranch income statement to reflect changes in resource use is called *budgeting*. Budgeting has been defined as a procedure to estimate the probable effects of production changes on costs and returns (Caton, 1957).

More specifically for our example and for economic analysis of range improvements in general, the required procedure used is called *partial budgeting* (Subcommittee on Range Research Methods of the Agricultural Board, 1962). As implied by the name, partial budgeting involves the economic analysis of a small change in resource use (such as a range improvement) and does not require a completely revised ranch income statement. Many items such as taxes, utilities, labor costs, and real estate capital costs remain unchanged. Only those items actually affected by a range improvement project need to be accounted for in the partial budgeting procedure. Table 9.6 represents the revenue side of partial budgeting; the cost side of the process appears in Table 9.7.

Budgeting in range economics research has given way, in recent years, to linear programming (LP) and other optimization techniques of planning investments in range improvements or practices. In contrast to partial budgeting, which deals with only one proposed project or practice at a time, LP seeks the optimum combination of proposed range management practices (Jameson et al., 1974). LP has been used both to estimate the minimum cost production process for a specific output and to estimate the maximum profit allocation of limited factors of production among alternative enterprises. Detailed discussions of the mathematics and computational procedures are readily available from several sources (Kim, 1971; Beneke and Winterboer, 1973; Jameson et al., 1974) and an example will not be included here.

Examples of LP applications to range research and management problems include Sharp and Boykin (1967), D'Aquino (1974), Bartlett et al. (1974),

Table 9.7 Hypothetical Costs of a Prescribed Burning Project, 1980

Initial treatment cost	
417 acres @$8	$ 3,336.00
Induced investment	
20 cows @$426	8,520.00
3 heifers @$320	960.00
1 bull @$1000	1,000.00
Total added investment	$10,480.00
Added annual cash costs	
Hay (reduced hay sales)	$ 2,200.00
Veterinary	33.33
Bull replacement	333.33
Other	133.40
Total added annual cash costs	$ 2,700.00

Leistritz and Qualey (1975), Child and Evans (1976), Hewlett and Workman (1978), and Torell et al. (1982). Various modifications of LP have been used on a wide variety of range management problems. Hunter et al. (1976) examined optimal stocking level with chance-constrained programming. Bartlett and Clawson (1978) used goal programming to compare the objectives of profit, meat production, and efficient use of energy. Grazing system impacts on risk and stability of ranch income were examined by Whitson et al. (1982). Each of these sophisticated computer techniques are based on the procedures of partial budgeting. The computer only serves to make the process easier and faster and to allow numerous production constraints to be analyzed simultaneously.

Expected Project Life

Reinvasion of undesirable shrubs or trees into a burned, seeded, sprayed, or mechanically treated range area is a gradual process. While increased forage production and improved livestock carrying capacity do not end abruptly, a point in time will be reached when deterioration is sufficient to end the effective life of the improvement project. Estimates of the duration of range improvements is subjective at best. Still, such estimates must be made in economic evaluations of range improvements and they should be based on research results and practical management experience in the vegetation type in which a range improvement project is being considered.

Seedings of improved grass species that have remained productive for 30 years or more are not unusual. Even less drastic treatments, such as prescribed burning in the sagebrush-grass type in Idaho, have shown impressive increases in carrying capacity that prevailed in excess of 15 years (Pechanec et al., 1954). It should be remembered that rate of reinvasion depends primarily on the density of plants surviving treatment (Johnson and Payne, 1968) and the intensity and timing of grazing use following treatment (Pechanec et al., 1954). Range improvements involving vegetation conversion are often implemented to alleviate seasonal shortages of forage. Especially in the case of range seedings, the seasonal forage produced is intended for intensive use during seasons of the year that may be detrimental to the health of the seeded plants. These facts should be borne in mind when estimates of effective project life are being made. For the analysis to follow, a conservative life of 15 years will be assigned to our prescribed burning example.

Expected Project Costs

The largest component of range improvement costs is usually the original investment in the actual land treatment or facility construction. Included are costs of machine operation, seed, spray, construction materials, labor, supervision, transportation, and planning. The magnitude of these costs may vary widely depending on (1) whether machinery is owned by the rancher or hired on a contract basis and (2) whether necessary labor is hired or supplied by the rancher and his family.

Also important in the determination of economic feasibility are *induced costs*. These costs are not part of the original treatment or construction but are incurred as a direct result of project implementation. Included are such items as construction and maintenance of fences and stock water facilities necessary for grazing deferment of treated areas. Induced costs can be an important factor in the economic success of improvement projects. In big sagebrush control for example, Kearl and Brannan (1967) reported that variation in fencing requirements may cause a threefold difference in per acre treatment costs of small tracts of land.

Another important induced cost associated with treatments such as seeding, burning, and spraying is that of the grazing deferment itself. If unrestricted, both livestock and big game animals often concentrate on recently treated areas, due apparently to increased forage palatability and improved availability (Vallentine, 1971). Required deferment may represent a significant cost. Pechanec et al. (1954), for example, recommended that when surviving native grasses are being relied upon to revegetate treated areas, burns in the sagebrush-grass type should be completely rested for one year following treatment and grazed at only 25 percent of posttreatment capacity the next year. Deferment costs are perhaps most easily expressed in terms of private lease rates for alternative sources of range forage (Nielsen and Hinckley, 1975).

Forage Value Based on Private Lease Rate The first of the two alternative approaches to pricing project benefits consists of valuing AUM produced by the prescribed burning directly at the private market lease rate. For the purpose of illustrating the first approach, project costs are set at $8 per acre. This figure represents the original investment in the burning treatment (firebreaks, ignition fuel, ignition labor and equipment, and standby fire suppression labor and equipment) for the situation where no additional fences or stock water facilities are required. For the sake of simplicity in our example, the induced costs of grazing deferment are not included in the calculations. Deferment costs cause the flow of net returns from the burning project to be nonuniform, but discussion of the analytical technique required to incorporate these important costs is postponed until Appendix II of this chapter.

Forage Value Based on Livestock Production The second technique of pricing project benefits consists of valuing increased forage in terms of resulting increases in cattle production. This second pricing approach requires accounting for several induced investment and annual operating costs of the burning treatment. In addition to the $8 per acre initial cost of the burning treatment itself, an expanded cow herd is needed to take advantage of the increased forage production.[16] Additional cows and heifers will likely be bought through increased retention of heifer calves at sale time. Although this allows the rancher to avoid a cash purchase of additional animals, cash income foregone by not selling these heifers and the feed costs required to bring them to reproductive maturity are real costs and must be included as part of the original range improvement investment. The easiest, and perhaps the most accurate, way of assigning these costs is simply to value the expanded herd at their current market price. As shown in Table 9.7, the 20 cows were valued at $426 each and the three replacement heifers at $320 each (1980 prices from Chapter 2).

Several increases in annual operating costs are also induced by the burning project. The easiest way to identify the additional costs associated with the 20 additional cows made possible by the prescribed burn is to carefully review the cost entries in Table 2.3. Feed costs will increase (actually crop returns will decrease) by $2200 due to feeding 55 additional tons of hay (138 additional AUM fed November–April)[17] valued at the 1980 price of $40 per ton. This added annual cost is entered in Table 9.7. Hired labor and repairs are not changed by the cow herd increase but veterinary expenses can be expected to increase by the average cost per cow, $500/300 = $1.67 for each cow added or a total of $33.33. Due to the $1000 additional bull required and the three-year retention in the herd, additional bull replacement costs average $333.33 per year. The "Other" cost category can also be expected to increase by the herd average for each of the 20 added cows or $133.40 additional annual costs.

Because the project has already been assigned a finite life of 15 years and the IRR calculations to follow are based on that definite expectation, depreciation costs of both the actual burning treatment and the expanded cow and heifer herd are automatically taken into account.[18] Since bull depreciation is accounted for as a cash replacement cost, the depreciation figures of Table 2.3 are not changed by the range improvement.

Interest Rate on Borrowed Capital

A crucial factor in any economic analysis of proposed range improvements (private or public) is the interest rate to be used for either (1) the discounting calculations underlying the PNW or B/C criterion or (2) comparison with the IRR generated by the project.

Ideally, the interest rate used for discounting or comparison should be the higher of either the interest rate that the investor must pay for borrowed capital or the opportunity cost rate (the rate of interest that could be earned if the capital required for the range improvement project were instead invested in its highest yielding alternative use). The best alternative use of range improvement capital is not always known. For this reason, and in an attempt to provide more widely applicable management recommendations from range economics research, a representative borrowing rate is often used as the interest rate for discounting or comparison purposes. Since the expected prescribed burning returns are measured in real (noninflated) prices, the appropriate borrowing or opportunity cost rate for this example is a real rate of interest.

ANALYSIS AND INTERPRETATION

All information required for economic analysis of our prescribed burning example is now available. Analysis and interpretation will now be demonstrated for both techniques of pricing range improvement benefits: (1) valuation of range forage based on the private market rate for leased forage and (2) valuation of range forage based on the value of increased livestock production.

Forage Value Based on Private Lease Rate

Priced at the 1980 private range lease rate of $8.00 per AUM (USDA, 1980), the 0.30 AUM per acre increase expected annually from the prescribed burn will yield a per acre net annual revenue (R) of $2.40 during the next 15 years (n). Since the following discounting calculations are based on the premise that the project will be completely used up in 15 years, depreciation is automatically taken into account and there is no need to subtract a depreciation allowance from R. Substituting these figures, along with the $8.00 initial investment per acre (I), into the IRR equation we obtain

$$\frac{I}{R} = \text{(Present Worth of One Per Period)},$$

$$\frac{\$8}{\$2.40} = 3.333 = \text{(Present Worth of One Per Period, 15 years)}.$$

Inspection of the 15-year row of Table 8.5 reveals that 3.333, our calculated Present Worth of One Per Period value, corresponds to an IRR far in excess of fourteen percent, the highest interest rate printed in Table 8.5. Referring to tables by Gushee (1968), we find that the IRR generated by our prescribed burning example is about 29 percent.[19] This rate of return can be compared directly with the real rate of interest paid on funds borrowed to implement range improvements. Interpretation of the rate calculated for our example is that the sagebrush burning is economically feasible if two conditions are met: (1) Range improvement funds can be borrowed at a real interest rate not exceeding 29 percent and (2) no alternative range improvement practice (or other investment opportunity) promises a real return greater than 29 percent.

Interpretation of calculated IRRs might be further clarified by referring to some actual rates of return for sagebrush control (Table 9.8) reported by Nielsen and Hinckley (1975). Since real (inflation-free) bank interest rates for range improvement loans would currently be in the neighborhood of 4 percent, the 23.85 percent IRR for the prescribed burning project easily meets the first condition that IRR exceed the borrowing rate. As indicated in Table 9.8, sagebrush burning also satisfies the second condition that IRR must exceed the opportunity cost rate. Burning promises the highest IRR of any sagebrush

Table 9.8 Internal Rates of Return for Several Big Sagebrush Control Methods and Degrees of Grazing Nonuse

Control method and type of deferment	IRR (%)
Prescribed burning	
Nonuse: rest for one year, graze at 25 percent of potential during second year	23.85
Chemical spraying	
Nonuse: rest for two years	15.62
Nonuse: graze at 50 percent of potential for two years	20.00
Chaining	
Nonuse: none	11.83
Plowing and seeding	
Nonuse: rest for two years	7.50
Rotobeating	
Nonuse: rest for one year	0.00

Source: Nielsen and Hinckley (1975).

conversion practice. Thus the rancher can afford to borrow the funds necessary to implement prescribed burning and this practice represents the best use of capital for improvement of sagebrush range.

Factors Affecting Economic Feasibility As indicated in the sections above, numerous factors have direct bearing on the economic success of range improvement projects. Included are: (1) magnitude of herbage increase; (2) percentage of herbage increase that can be safely harvested as forage; (3) value of each additional AUM of forage produced; (4) productive life of the project; (5) initial investment required, influenced by such factors as whether equipment is owned or contracted and whether hired or family labor is used; (6) induced costs such as fences, water facilities and grazing deferment; (7) the borrowing or opportunity cost rate.

Site Selection One additional factor influencing the economic success of range improvements deserves mention here. As emphasized by Nielsen and Hinckley (1975), an important determinant of the success of vegetation manipulation projects is selection of the site to be treated. As they point out, the vigor of the vegetation presently occupying a range site is a good index of site potential. Tall, robust big sagebrush, for example, likely indicates a deep, well drained soil receiving sufficient moisture to support a productive stand of native grasses.

Table 9.9 Hypothetical Per Acre Forage Production Increases, Required Investment, and Annual Returns to Prescribed Burning on Two Range Sites

	Site A	Site B
Pretreatment air-dry usable forage production (lb / acre)	66	99
Posttreatment air-dry usable forage production (lb / acre)	132	198
Increased air-dry usable forage production (lb / acre)	66	99
Increased carrying capacity AUM / acre (660 lb air-dry per AUM)	0.10	0.15
Per acre value of increase at $8 per AUM	$ .80	$1.20
Per acre cost of burning	$8.00	$8.00
Life of project (years)	15	15
Internal rate of return (%)	5.56	12.40

Source: Workman (1976).

Another observation by Nielsen and Hinckley (1975) relative to site selection is that a direct relationship exists between pretreatment and posttreatment forage production. The hypothetical example of Table 9.9 will be used to demonstrate the importance of careful selection of areas to be treated. Suppose our alternatives for range burning consist of two sites, A and B, both 1000 acres in size (Table 9.9). Suppose further that each site is currently in fair condition and that application of a well planned prescribed burn will improve range condition of both sites to good and simultaneously double production of usable forage. Although treatment doubles forage production on each site, treatment costs are no higher on the high potential area than on the low potential site. Also, while pretreatment production on site B is only 50 percent higher than on site A, improvement of site B yields a return more than double the IRR obtained on site A. The conclusion is simple but important: If our goal is to maximize AUM increases from a limited range improvement budget, we should treat our best sites first. In sharp contrast to this simple rule, range managers have often concentrated improvement efforts on the worst sites first because the poor sites "need" improvement more than the good sites. This philosophy is based on confusion between what we would like to do and what we can afford to do and can cause serious errors in the allocation of range improvements between sites.

Forage Value Based on Livestock Production

All information needed to evaluate the economic feasibility of the prescribed burning project in terms of increased livestock production is available in Tables 9.6 and 9.7. Combining this information in Table 9.10, we can now complete the analysis. Substitution of the $10,480 required investment (I) and

Table 9.10 Hypothetical Values Derived for a Range Improvement Resulting in 20 Additional Cows, 1980

Added cash returns ($)	4,132.00
Added cash costs ($)	2,700.00
Added net ranch income ($)	1,432.00
Added initial investment ($)	10,480.00

$$\frac{I}{R} = \frac{[1 - (1+i)^{-n}]}{i}$$

$$\frac{10,480}{1432} = 7.318 = \text{(Present Worth of One Per Period, 15 years)}$$

IRR = 10.68%

---

the $1432 net revenue ($R$) into our IRR equation yields a Present Worth of One Per Period value of 7.318. Examination of the 15-year row of Table 8.5 shows that the IRR is just over 10 percent (use of a financial calculator gave a real IRR of 10.68 percent).

If the increased forage produced by the prescribed burn is to be used to expand the breeding herd, the 10.68 percent IRR is the interest rate to be directly compared with the real cost of borrowed capital and the real rate of return promised by all other potential investments.[20] It should be emphasized that this is the relevant rate for analysis of the range improvement despite the fact that the nominal rates of return on total owned ranch capital are only slightly improved as shown by the before and after comparisons of Table 9.11.

Table 9.11 consists of the cost, return, and investment information from Table 2.4 (Before) and the same information revised to incorporate all changes brought about by the prescribed burn and resulting herd expansion (After). Perhaps the most relevant change from the rancher's viewpoint is the $54 increase in annual net return available for family living expenses. As pointed out by Nielsen (1967), the small rancher is often more concerned with improving the standard of living of his family than with increasing his net worth. Consequently, the small ranch operator will not (and cannot) invest in range improvement projects that promise attractive future returns but reduce current net family income. The $54 increase in annual net return available for family living expenses is the amount available after servicing a 1980 loan of the entire initial investment of $10,480 for 15 years at 10 percent.

No change in land appreciation is shown in Table 9.11. Unfortunately, it is doubtful that the prescribed burn would have any appreciable effect on market value of the ranch. The $330 increase in payment toward mortgage principal is the first year principal payment associated with the $10,480 range improvement loan. Both the original 8.18 percent nominal return and the 8.23 percent return after improvement were calculated using $733,164 as the value of owned ranch capital.

It should be emphasized once more that the economic merits of this range improvement project should not be judged by the small increase in nominal

Table 9.11 Modified Income Statement for a Hypothetical Utah Cattle Ranch[a]

	Before improvement	After improvement
Annual cash returns	$70,000	$74,132
Annual cash costs	−31,000	−33,700
Depreciation costs	−7,000	−7,000
Net ranch income	32,000	33,432
Loan service cost		
Real estate	5,909	5,909
Working capital	13,715	15,093
Total	−19,624	−21,002
Net return available for family living expenses	12,376	12,430
Land appreciation	+51,300	+51,300
Payment to mortgage principal	+11,290	+11,620
Gross proceeds to ranch investment	74,966	75,350
Value of operator and family labor and management	−15,000	−15,000
Net proceeds to owned ranch capital	59,966	60,350
Percent return on $733,164 owned ranch capital	8.18	8.23

[a]Revised to include 20 additional cows made possible by 125 additional AUM from range improvement, 1980.

rate of return on owned ranch capital. From the viewpoint of the ranch family that intends to continue in the livestock business anyway, the 10.68 percent real return expected from the prescribed burn is still the relevant measure of the economic feasibility of the project. In fact, even if a *negative* percent return on owned ranch capital resulted both before and after range improvement, any decreases in such losses can be viewed as positive returns by the rancher who is committed to continuing his operation. In our hypothetical example, the 10.68 percent real return on the additional investment in the livestock operation may well be higher than that promised by any other investment opportunity available to the rancher committed to retaining ownership of land and livestock.

APPENDIX I Disagreement Among Investment Criteria — A Proposed Solution

A Review of the Problem

The question of which of the three standard criteria (IRR, B/C, and PNW) to use in evaluating investment projects has long been a source of controversy among economists (Gardner 1963; LeBaron 1963) but one that only recently has received much attention from range managers. The recent interest of the range profession in the choice of investment criteria has been triggered primarily by modern computer programs such as the U.S. Forest Service Invest III investment program that automatically generates measures of all three criteria from raw data dealing with range improvement investment, life expectancy, expected productivity response, etc. The question of which criterion to rely on and which to disregard has come about simply because the three criteria, as commonly calculated, produce contradictory results. The purpose of this appendix is to demonstrate that if the three criteria are correctly calculated, each gives the same correct investment decision.

The controversy concerning which investment criterion is the correct standard for selecting from among available investment alternatives was well summarized over 20 years ago by Gardner (1963) and LeBaron (1963). Gardner confined his discussion to the choice between IRR and PNW, while LeBaron also included B/C, the criterion most commonly used by federal land management agencies. A brief review of the conclusions of these two authors follows.

Gardner (1963) began by pointing out the wide variability in rates of return reported for rangeland reseeding projects in different parts of the intermountain area (Caton et al., 1960; Lloyd and Cook, 1960; Caton and Beringer, 1960; Gardner, 1961). He then demonstrated that most of the apparent variation in reported rates was due to differences in methods used to calculate (what was then commonly called) the rate of return rather than to true differences in costs and returns. Based on his review of several classical treatments of the problem of capital budgeting (Dean 1964; Alchian 1955; Lorie and Savage, 1955), Gardner recommended that IRR, rather than PNW, be used as the criterion for ranking range improvement projects that are mutually exclusive due to limited investment funds. Advantages claimed for IRR were (1) the calculated rate is directly comparable with the compound interest rate paid for borrowed capital, (2) it is not necessary to undertake the difficult task of selecting the "correct" interest rate for PNW discounting calculations, and (3) IRR standardizes projects with respect to size and expected life. Gardner did caution, however, that the listed advantages are based on the assumption that net cash flows to a short-lived project can be reinvested at the IRR generated by the project to give a useful life equal to the longest-lived project under consideration.

LeBaron (1963) provided a rebuttal to Gardner's claim of superiority of IRR over PNW. Drawing on Hirshleifer et al. (1960), he criticized the project rankings of IRR as inconsistent with those of PNW whenever project lives differ and net returns cannot be "immediately and perpetually reinvested at their own internal rates of return" (Alchian, 1955). By posing the following questions, LeBaron also took issue with the third advantage claimed by Gardner (and by Solomon, 1956) that IRR standardizes projects with respect to size:

> Suppose mutual exclusiveness, what does it mean to say that average internal rate of return is 9 percent on $40,000 or 8 percent on $50,000? Which is best (assuming equal lives)? Can the decision be made without the introduction of some side calculations (concerning, say, the $10,000 difference that might be invested in the market, etc.?

Some Range Improvement Examples

Disagreement Among Criteria Some simple range improvement examples may clarify the above concerns and help point the way to a workable solution to the problem of disagreement among investment criteria. Suppose a rancher has two range

Table 9.12 Comparison of Two Hypothetical Range Improvement Projects by Three Investment Criteria, Standard Calculation Method.[a]

	Project	
	Spraying sagebrush	Developing stock water
Project life (years)	20	20
Annual net return ($)	142	320
Initial investment ($)	1000	2400
PNW at 10% discount rate ($)	1209 − 1000 = 209	2724 − 2400 = 324
IRR (%)	12.96	11.93
B / C at 10% discount rate	$\frac{1209}{1000} = 1.21$	$\frac{2724}{2400} = 1.14$

Source: Workman (1981).
[a] Equal lives and different required investments

improvement projects that appear promising for his operation (Table 9.12). Suppose further that the rancher can borrow up to $2400 at 9 percent real interest and his best alternative investment opportunity promises to yield 10 percent. The relevant interest rate to be used in discounting future project net returns back to the present is, of course, the higher of the borrowing rate and the opportunity cost rate or, in this case, 10 percent. Thus, for the spraying project the present value of $142 real net return to be received annually for 20 years is $142 × 8.5136 = $1209 where 8.5136 is the Present Worth of One Per Period factor for 20 years and 10 percent.

The rancher's problem is one of deciding which range improvement, if either, to invest in. As shown in Table 9.12, both range improvement projects appear economically feasible by the standard calculation method as measured by all three investment criteria. PNW indicates that both projects yield a net profit, IRR[1] shows that both yield an annually compounded rate of interest greater than the 9 percent cost of borrowing and the 10 percent opportunity cost, and B/C reveals that each $1 invested produces more than $1 in returns.

Although both improvement projects have proved to be economically feasible by each of the three investment criteria, the rancher must decide which of the two projects to implement. Since his banker has agreed to lend a maximum of only $2400, the rancher clearly does not have enough capital to invest in both projects. As shown in Table 9.12, the investment criteria disagree as to which of the two projects is the better investment. Both IRR and B/C select the spraying project, while PNW indicates that the stock water development is the better investment.

The rancher's goal is, of course, to maximize the present value of the net return (PNW) from investing the total $2400 of available capital. Since a net return of $324 is obviously preferred to one of $209, it appears that the PNW criterion is correct in choosing the stockwater development, while the other two criteria have both selected the wrong project. Further discussion of this point follows.

Now suppose the rancher is faced with choosing between two entirely different projects—a prescribed burn and the implementation of a grazing system (Table 9.13). He is now able to borrow up to $10,000 at 9 percent interest, and his opportunity cost remains at 10 percent. Again, by the standard calculation method, there is disagreement among the three investment criteria. At a discount rate of 10 percent, both PNW and

Table 9.13 Comparison of Two Hypothetical Range Improvement Projects by Three Investment Criteria, Standard Calculation Method, 10 and 10.5% discount rates[a]

	Project	
	Prescribed burning	Grazing system
Project life (years)	15	30
Annual net return ($)	1391	1132
Initial investment ($)	10,000	10,000
PNW at 10% discount rate ($)	10,580 − 10,000 = 580	10,671 − 10,000 = 671
B/C at 10% discount rate	$\frac{10,580}{10,000} = 1.06$	$\frac{10,671}{10,000} = 1.07$
IRR (%)	11.00	10.80
PNW at 10.5% discount rate ($)	10,285 − 10,000 = 285	10,242 − 10,000 = 242
B/C at 10.5% discount rate	$\frac{10,285}{10,000} = 1.03$	$\frac{10,242}{10,000} = 1.02$
IRR (%)	11.00	10.80

Source: Workman (1981).
[a] Different lives and equal required investments

B/C select the grazing system as the better project, which seems appropriate since the rancher logically prefers a profit of $671 to one of $580. But the arithmetic of Table 9.13 is based on a 15-year project life for the prescribed burn and a 30-year life for the grazing system. If there is an opportunity to reinvest the funds recaptured from the 15-year reseeding project at 11 percent, then of course the rancher would prefer a rate of return of 11 percent on a $10,000 investment to one of 10.8 percent on the same amount of capital. Thus, IRR appears to have made the correct choice, while both PNW and B/C have made the wrong selection. It seems, then, that the standard calculation method leaves room for honest confusion as to which of the two projects is truly superior. We will return to this question later.

Inconsistent Criteria As mentioned previously in the review of Gardner's (1963) paper, the fact that it is not necessary to select a discount rate for IRR calculations is often claimed as an advantage for this criterion. As commonly calculated,[2] IRR is not dependent upon the borrowing or opportunity cost rate and thus consistently selects the same project from a choice of two or more investment opportunities, while project selection by both PNW and B/C depends upon the interest rate used for discounting, causing the two latter criteria to be labelled "inconsistent" (Workman, 1976). Table 9.13 illustrates the inconsistency. Using a 10 percent discount rate, both PNW and B/C choose the grazing system, while the prescribed burn is favored by IRR. Changing the discount rate to 10.5 percent and leaving all other Table 9.13 numbers the same causes both PNW and B/C to reverse their original selections and choose the prescribed burn as the superior project. Since standard calculations of IRR are not dependent upon the discount rate, it consistently selects the burning project.

ECONOMIC ANALYSIS OF PRIVATE RANGE IMPROVEMENTS

Normalization—A Solution

Despite the above confusion, if the three criteria are correctly calculated all will select the same investment project in all cases. Both the contradictory project selections and the resulting confusion as to which criterion to follow are due to several crucial items of information being left out of the calculations. Although virtually all benefit–cost calculations in natural resource periodicals, research bulletins, environmental impact statements, and textbooks have been done as previously shown, such calculations are incorrect. As explained by Mishan (1976), in order for investment alternatives to be correctly (and consistently) compared by any of the three common criteria, they must be adjusted or normalized to take the following items into account:

1. Differences in project size (required initial investment).
2. Differences in project expected life.
3. Rate of return on best alternative investment (opportunity cost).

A normalization procedure will now be demonstrated that corrects for the items listed above and, thus, eliminates project selection disagreements among investment criteria.

Correcting for Project Size Differences The key to correct calculation of parameters for the three criteria is explicit recognition of the rate of return available on the best alternative use of capital (initial capital investment as well as reinvestment of annual project returns). In Table 9.12, the best alternative to investing in sagebrush spraying and stock water development promised a 10 percent rate of return. Recognition of this opportunity cost rate allows adjustment to be made for differences in project size. The required initial investment for the stock water development consists of the entire $2400 available, while the required investment for the spraying amounts to only $1000. However, the additional $1400 not required for the spraying project can be invested in the best alternative investment at 10 percent interest Thus, as shown in Table 9.14, the total future value of the spraying project at the end of year 20 consists of the $142 annual return from the $1000 investment, compounded (Future Worth of One Per Period factor) at 10 percent over the 20-year life (since the annual net return can be reinvested at the opportunity cost rate) ($8133) plus the future value of the excess $1400 compounded (Future Worth of One factor) forward to the end of year 20 (since this amount can also be invested at the opportunity cost rate) ($9419) or a total value of $17,552. Similar calculations yield a total future value for the stock water project of $18,328 (Table 9.14).

Now that the investment and return data for the two projects have been adjusted for differences in project size, parameters for the three investment criteria can be correctly calculated. As mentioned above, the rancher's investment goal is to maximize the present net worth[3] resulting from investing the entire $2400 available. For the spraying project PNW is the present value (Present Worth of One factor) of the $17,552, to be received at the end of 20 years ($2609) minus the $2400 originally invested or $209. Similarly, PNW[4] for the stock water project is the present value of the $18,328 future value ($2724) minus the $2400 initial investment or $324. Since the stock water promises a considerably higher PNW, it is the preferred project as measured by PNW. But what about the other two criteria? As shown in Table 9.14,

Table 9.14 Comparison of Two Hypothetical Range Improvement Projects by Three Investment Criteria, Normalized Calculation Method[a]

	Spraying			Stock water	
	Year 0	Year 20	Year 0		Year 20
	−$1000	$142 annually for 20 years at 10% → $8133	−$2400	$320 annually for 20 years at 10% →	$18,328
	−$1400 −$2400	$1400 at 10% for 20 years → 9419 $17,552	−$2400		$18,328
	+$2609 $ 209	10% for 20 years → $17,552	+$2724 $ 324	10% for 20 years →	$18,328
PNW:	$209		PNW:	$324	
B/C:	$2609/$2400 = 1.09		B/C:	$2724/$2400 = 1.14	
IRR:	$2400 —— 20 years at i% —— $17,552		IRR:	$2400 —— 20 years at i% ——	$18,328
	IRR = i = 10.46%		IRR = i = 10.70%		

Source: Workman (1981).
[a] Equal lives and different required investments

Table 9.15 Comparison of Two Hypothetical Range Improvement Projects by Three Investment Criteria, Normalized Calculation Method[a]

	Prescribed burn			Grazing System		
	Year 0	Year 15	Year 30	Year 0		Year 30
	−$10,000	$1391 annually → $44,196 $\xrightarrow{10\%}{15\text{ years}}$	$184,618	−$10,000	$1132 annually → $\xrightarrow{10\%}{30\text{ years, }10\%}$	$186,207
		15 years, 10%			30 years, 10%	
	−$10,000	10% for 30 years	$184,618	−$10,000	10% for 30 years	$186,207
	+$10,580			+$10,671		
PNW:	$580			$671		
B/C:	$\dfrac{\$10,580}{\$10,000} = 1.06$			$\dfrac{\$10,671}{\$10,000} = 1.07$		
IRR:	$10,000 $\xrightarrow{\text{30 years}}{\text{at }i\%}$		$184,618	$10,000 $\xrightarrow{\text{30 years}}{\text{at }i\%}$		$186,207
IRR $= i = 10.21\%$				IRR $= i = 10.24\%$		

Source: Workman (1981).

[a] Different lives and equal required investments, 10 percent discount rate.

normalized B/C and IRR[5] also select the stock water project. Thus, when correctly calculated, each of the three investment criteria chooses the same project.

Correcting for Project Life Differences As mentioned in the preceding section, the disagreement among the three investment criteria in Table 9.13 is due to the difference in expected lives of the two projects. As with the normalizing procedure used to correct for project size differences, the key to correcting for differences in project lives is recognition of the rate of return available on the best alternative use of capital (the opportunity cost). As already established, the best alternative to the two projects considered in Table 9.13 promises a 10 percent return. This information allows the life of the shorter-lived project (prescribed burn) to be adjusted to equal that of the longer-lived project (grazing system). This is done by calculating the future value (at the end of year 15, the final year of the prescribed burn) of the $1391 annual return and then simply compounding this value ($44,196) forward at 10 percent for 15 years to give the future value of the prescribed burn income stream at the end of year 30 ($184,618) (Table 9.15). By similar calculations, the future value of the grazing system is $186,207. Based on these adjustments, correct values for IRR and B/C can now be calculated. Since normalization of B/C and PNW for nonuniform lives consists of first compounding returns forward and then discounting them backward at the same interest rate (the opportunity cost), these two criteria are already correct in Table 9.13 and are not altered by the normalization procedures of Table 9.15. Again, the rancher's goal is to maximize PNW resulting from investing the total available capital ($10,000). Discounting future values of the two projects back over 30 years at 10 percent and subtracting the $10,000 initial investment required by each yields PNW values of $580 and $671, respectively, for the prescribed burn and grazing system. Thus at a discount rate of 10 percent, the grazing system is the project favored by PNW and B/C just as in Table 9.13. But what about IRR, which chose the prescribed burn project in Table 9.13? After correction for the difference in project lives, the three criteria no longer disagree and IRR also selects the grazing system as the superior project.

Inconsistent Criteria What if the discount rate were increased from 10 to 10.5 percent as in Table 9.13? As shown in Table 9.16, the increase in discount rate causes each of the three criteria to now favor the prescribed burn over the grazing system. While the adjustment for the difference in project expected life ensures that the three criteria will consistently agree as to which project should be implemented, all three criteria are dependent upon the interest rate used for compounding and discounting and might be termed inconsistent with regard to this dependence.

Choosing between Two Projects—A Summary As demonstrated in Tables 9.14, 9.15, and 9.16, project normalization to correct for differences in project life and project size ensures that each of the three criteria results in the same correct choice between two range improvement projects. Normalization of both project life and project size is essential for correct project selection by IRR, while B/C requires normalization of only project size. PNW makes the same correct selection whether normalized or not. PNW values are unaffected[6] by normalization simply because the correction procedure involves the compounding forward and discounting backward of project returns at the same interest rate, the opportunity cost rate. In the interest of simplicity and ease of calculation, then, PNW is the criterion recommended for selection between two projects.

Table 9.16 Comparison of Two Hypothetical Range Improvement Projects by Three Investment Criteria, Normalized Calculation Method[a]

	Prescribed burn			Grazing system		
Year 0	Year 15		Year 30	Year 0		Year 30
−$10,000	$1391 annually 15 years, 10.5% → $45,987	$\dfrac{10.5\%}{15 \text{ years}}$ →	$205,622	−$10,000	$1132 annually 30 years, 10.5% →	$204,758

−$10,000
+$10,285   10.5% for 30 years $205,622

PNW: $285

B/C: $\dfrac{\$10,285}{\$10,000} = 1.03$

IRR: $\$10,000 \xrightarrow{30 \text{ years at } i\%} \$205,622$

IRR = i = 10.60%

−$10,000
+$10,242   10.5% for 30 years $204,758

PNW: $242

B/C: $\dfrac{\$10,242}{\$10,000} = 1.02$

IRR: $\$10,000 \xrightarrow{30 \text{ years at } i\%} \$204,758$

IRR = i = 10.59%

Source: Workman (1981).
[a] Different lives and equal required investments, 10.5 percent discount rate.

Table 9.17 PNW and IRR Values for Six Hypothetical Range Improvement Projects, Ranked by the B/C Criterion

Project	Required investment ($)	Annual net return ($)	Project life (years)	PNW ($, 10%)	IRR (%)	B/C (10%)
1. Spraying	1,000	142	20	209	12.96	1.21
2. Stock water	2,400	320	20	324	11.93	1.14
3. Cross-fencing	3,400	408	30	446	11.55	1.13
4. Grazing system	10,000	1,132	30	671	10.80	1.07
5. Prescribed burn	10,000	1,391	15	580	11.00	1.06
6. Seeding	5,000	508	30	−211	9.49	0.96

Source: Workman (1981).

Selecting an Optimum Combination from Three or More Projects

Often range improvement investment decisions consist of selecting several projects from among numerous promising alternatives In this case, the manager's goal is to maximize total net returns (PNW) produced by just exhausting all available capital. As might be expected, the selection procedure is more complicated than choosing between only two potential projects. For the simple (and rare) situation where capital is unlimited, the procedure involves investing in all projects promising a PNW value greater than zero (or B/C greater than one or IRR greater than the borrowing cost and opportunity cost rates). But in virtually all management situations, capital is limited and an optimum combination of projects must be selected that maximizes PNW while staying within the limited budget.

The Traditional Approach Suppose our borrowing cost and opportunity cost rates are both 10 percent and our investment problem is to select a combination of range improvement projects from those listed in Table 9.17 that will maximize PNW while not exceeding a specific budget constraint. Traditional procedure (Howe, 1971) has been to rank the potential projects by IRR (or B/C) values. Next, projects are selected by descending rank until the improvement budget is exhausted. It follows from the preceding discussion concerning the choice between two projects by nonnormalized IRR and B/C values that the traditional procedure can lead to incorrect investment decisions. In the following section, this fact is demonstrated and a method of avoiding project selection errors is described.

A Recommended Approach Fortunately, our goal in this case is to select a set of projects of various sizes that will maximize PNW while just exhausting available capital. Thus it is not necessary to normalize the projects for differences in required investment. However, correct ranking of the projects in Table 9.17 by IRR would require normalization for differences in project life. For this reason, the procedure recommended (and demonstrated) is as follows: (1) Rank the projects by B/C. (2) Select best projects first until the available capital is exhausted. (3) Perform PNW "side calculations" on the last two or three (or more) projects, as required, to verify the accuracy of the selected project combination.

ECONOMIC ANALYSIS OF PRIVATE RANGE IMPROVEMENTS 173

The six hypothetical range improvement projects in Table 9.17 have been ranked by descending B/C values and for comparison and verification purposes their PNW and IRR values are also shown. Once the projects have been ranked by B/C as shown, the decision as to which projects, if any, to implement is based on the amount of capital available. If, for example, total capital consisted of only $1000, the spraying project would obviously be selected, since it is the only alternative within reach of the limited budget.

If $2400 in capital were available, the investment decision would be more difficult. The B/C criterion (and IRR) favor the spraying over the stock water development. However, since (1) all values shown in Table 9.17 are nonnormalized, (2) the goal is to maximize PNW while staying within the budget constraint, and (3) the two projects are mutually exclusive (i.e., it is not possible to construct a smaller version of the $2400 stock water development in addition to the $1000 spraying project), we cannot rely on the B/C (or IRR) ranking and a PNW side calculation is required to test whether or not the spraying really is the superior project. As shown in Table 9.17, calculation of PNW values for both projects reveals that spraying yields only $209 while stock water development produces $324. Thus the choice of the spraying project by B/C (and IRR) is incorrect. Since investment in either project precludes investment in the other, the stock water project is clearly superior.

Now suppose $3400 in capital were available. Given this budget constraint, the B/C (and IRR) rankings select the spraying and stock water projects—a combination that exactly exhausts the available capital. In this case PNW side calculations for the two possible project combinations verify that spraying and stock water development represent the optimum investment combination. Together they yield a combined PNW of $533, while cross-fencing, the only alternative, produces a PNW of only $446 (Table 9.17). While in this example it is true that B/C (and IRR) did select the optimum investment combination, we were not sure it was the correct combination until it was verified by the PNW calculations.

Next suppose total investment capital amounted to $10,000. With this larger budget constraint, B/C (and IRR) select the first three projects (spraying, stock water, and cross-fencing). Although this three project combination requires a total investment of only $6800, it effectively exhausts the $10,000 budget because the projects are mutually exclusive (the remaining $3200 of unused capital is insufficient to implement any of the remaining three projects). Our necessary PNW side calculation shows that B/C (and IRR) have selected the correct project combination. A combined PNW of $979 is produced as compared with only $671 yielded by the next best alternative, the grazing system. Again it should be emphasized that we were not sure that B/C had selected the correct project combination until PNW values for the two investment alternatives were compared.

Suppose, finally, that the total budget amounted to $20,000. With this budget constraint the B/C ranking selects the first four projects (spraying, stock water, cross-fencing, and the grazing system). A PNW side calculation verifies this selection as the correct combination of projects (a combined PNW of $1650 versus only $1252 for a combination of the prescribed burn and the grazing system). It should be noted that in this case the IRR criterion makes the wrong selection (a combination of spraying, stock water, cross-fencing, and prescribed burning, resulting in a combined PNW of only $1559). As explained above, the IRR criterion provides a correct ranking only if projects are normalized for nonuniform lives (30 years for the grazing system versus 15 years for the prescribed burn). It is for this reason that B/C, rather than IRR, is recommended as the criterion to be used for initial ranking purposes when a selection must be made from among numerous potential range improvement projects.

It is worth mentioning that even if the range improvement budget were unlimited, only $26,800 could be efficiently utilized. Since the stated goal is to maximize net return from the total available budget, the seeding project in this example is clearly not economically feasible as measured by any of the three investment criteria and should not be implemented regardless of the amount of capital available.

From the examples above, it is apparent that when selecting an optimum investment combination from several possible projects, rankings by B/C (or IRR) are good first approximations and indispensable aids when confronted with long lists of potential projects. However, rankings by either of these criteria can be wrong (as evidenced by the $2400 budget constraint case above) and such rankings must be checked by PNW side calculation. Since the B/C criterion makes correct selections when expected project lives are nonuniform, B/C, rather than IRR, is recommended for use as a first approximation ranking device for numerous range improvement projects.

Selecting an Optimum Combination of Projects—A Summary When the management goal is to maximize net return from total available capital, the recommended procedure for selecting an optimum combination of projects from numerous potential investments is as follows: (1) Rank the projects in descending order of their B/C values. (2) Choose best projects first until the available capital is exhausted. (3) Perform PNW side calculations on the last two or three (or more) projects, as necessary, to verify the accuracy of the selected project combination.

APPENDIX II Modified IRR Calculations for Nonuniform Flows of Costs and Returns

The Concept

All of the range improvements discussed above are examples of the simple case where an initial investment yields an annuity or uniform annual net income flow. In reality, range improvements often produce nonuniform income streams. Annual income produced by a reseeding, for example, may increase during the first few years as the stand becomes established, then exhibit a gradual decline, and finally reach zero as undesirable plants reinvade the treated area. "Lumpy" maintenance and reinvestment expenditures, such a those required for necessary fences and stock water facilities, are another cause of nonuniform net income streams. While these nonuniform streams can be easily discounted in PNW and B/C calculations through the use of the Present Worth of One factors of Table 8.4, uneven flows greatly complicate determinations of IRR as commonly calculated.[1] A simple example will demonstrate why this is so. Suppose we are considering investing in the following range improvement:

Initial investment	Net return	
	Year 10	Year 20
$1000	$2000	$2000

Rather than a uniform annuity, this project promises to yield two lump sum payments at two future points in time. If we wish to compare this project with other potential investments, how can we calculate the IRR? In Chapter 8, we used Table 8.4 to calculate the rate of return generated by a single lump sum payment at some point in the future. We will now apply the same concept to determine the IRR generated by two future lump sums. We know that the column of Table 8.4 corresponding to the IRR we are seeking contains Present Worth of One (PWO) values for 10 years and 20 years such that the following is true:

$$\$1000 = PWO_{10}(\$2000) + PWO_{20}(\$2000).$$

What may at first seem to be an algebraically insolvable equation (two unknowns and only one equation) is actually quite easy to solve since we know that both of the PWO values we are searching for occur in the same interest rate column of Table 8.4. Let us begin with 8 percent:

$$0.4632(\$2000) + 0.2145(\$2000) = \$1201.16.$$

Since our calculated present value is greater than $1000 we know the IRR is larger than 8 percent. Referring to a more detailed PWO table such as supplied by Gushee (1968), let us next try 12 percent:

$$0.3220(\$2000) + 0.1037(\$2000) = \$851.40.$$

Now the calculated present value is less than $1000 so the generated IRR is somewhere between 8 and 12 percent. The PWO values for 10 percent yield a fairly close present worth (Gushee, 1968):

$$0.3855(\$2000) + 0.1486(\$2000) = \$1068.20.$$

Since the calculated present worth is still too large, the true IRR is somewhat larger than 10 percent. For most investment decisions, though, this estimate would be sufficiently accurate.

In this simple example a minimum of trial and error is involved in identifying the approximate IRR. IRR for even extremely complicated investments can be found with a few iterations, although the required arithmetic is very time consuming.[2]

An Application

Grazing deferment required for reseeding and vegetative conversion projects is one of the most common causes of nonuniform flows of net returns. The hypothetical prescribed burning project illustrated in Figure 9.1 will be used to demonstrate the technique of calculating the IRR for range improvements that require grazing deferment. The method used follows that of Nielsen and Hinckley (1975).

The calculated IRR of nine percent for the prescribed burn of Figure 9.1 was based on a treatment cost of $3.50 per acre and a deferment cost of $0.50 per acre during the year of treatment, for a total per acre initial investment of $4.00. Pretreatment carrying capacity was set at 0.1 AUM per acre and a doubling of carrying capacity was expected to result from the prescribed burn. Priced at $5 per AUM, this gave an expected net return of $0.50 per acre over the 15-year life of the project. What was not recognized in the original IRR of nine percent was that in addition to full deferment during the year treatment occurred, partial deferment is also required during the first year after treatment (year 1 in Figure 9.1). Following deferment recommendations of Pechanec *et al.* (1954), the treated area can be grazed at only 25 percent of posttreatment sustained yield during the first year after treatment. Thus the year 1 deferment cost consists of a gross loss of $0.50 per acre (0.1 AUM pretreatment carrying capacity times $5 per AUM) minus the actual return of $0.25 per acre (25 percent of $0.50 pretreatment return plus 25 percent of the $0.50 increase in return due to treatment) for a net loss (net difference between pretreatment and posttreatment yield) of $0.25 (Table 9.18). While seemingly insignificant, this small deferment cost will have a marked impact on the IRR yielded by the project and should be included in the analysis.

Calculation of the correct IRR generated by this project using the PWO values of Table 8.4 could be very time consuming. However, we know the true IRR must be less than the nine percent calculated when first year deferment is not included. We might try the eight percent PWO values from Table 8.4. Eight percent gives a present worth of the income stream shown in the second column of Table 9.18 of only $3.59. Since this is far short of the initial investment of $4.00 per acre, we might now try seven percent. But the $3.85 present worth at seven percent is still $0.15 too small and we next try six percent. This gives a present worth of $4.15 which is $0.15 too large and we know we have bracketed the true IRR. We next try 6.5 percent (Gushee, 1968), which gives the

Table 9.18 Calculation of IRR for a Nonuniform Income Stream Using Present Worth of One Values

Initial investment (per acre)			
Treatment			$3.50
Deferment during initial year			0.50
Total			$4.00

Present worth of net income stream (per acre)			
Year	Income ($)	PWO value (6.5%)	Present worth ($)
1	−0.25	0.9390	−0.23
2	0.50	0.8817	0.44
3	0.50	0.8278	0.41
4	0.50	0.7773	0.39
5	0.50	0.7299	0.36
6	0.50	0.6853	0.34
7	0.50	0.6435	0.32
8	0.50	0.6042	0.30
9	0.50	0.5673	0.28
10	0.50	0.5327	0.27
11	0.50	0.5002	0.25
12	0.50	0.4697	0.23
13	0.50	0.4410	0.22
14	0.50	0.4141	0.21
15	0.50	0.3888	0.19
Total			$3.98

results shown in Table 9.18 and our iterative process is complete. The total present worth of the income stream discounted at 6.5 percent is $3.98, almost equal to the initial investment of $4.

Although small, the $0.75 per acre first year difference in net income between a $0.50 net return and a $0.25 net loss has reduced the calculated IRR from 9 percent to 6.5 percent. Obviously the inflated IRR resulting from leaving deferment costs out of the analysis could lead to errors in range improvement investment decisions.

LITERATURE CITED

Alchian, A. A. 1955. The rate of interest, Fisher's return over cost, and Keynes' internal rate of return. *Amer. Econ. Rev.* 45:938-943.

American Institute of Real Estate Appraisers. 1972. *Real Estate Appraisal Course V. Grazing Lands and Cattle Ranches.* Logan, Utah, June 24-July 1. 160 pp. (Mimeo.).

Barkley, P. W., and D. W. Seckler. 1972. *Economic Growth and Environmental Decay.* Harcourt Brace Jovanovich, New York. 193 pp.

Bartlett, E. T. and W. J. Clawson. 1978. Profit, meat production, or efficient use of energy in ranching. *J. An. Sci.* 46:812-818.

Bartlett, E. T., G. R. Evans, and R. E. Bement. 1974. A serial optimization model for ranch management. *J. Range Manage.* 27:233-239.

Beneke, R. R. and R. Winterboer. 1973. *Linear Programming Applications to Agriculture.* Iowa State Univ. Press, Ames. 243 pp.

Caton, D. D. 1957. Budgeting in research relative to the use of range resources. In *Economic Research In the Use and Development of Range Resources. A Methodological Anthology.* Western Agricultural Economics Research Council Report No. 1, Berkeley, CA. 151 pp. (Mimeo.).

Caton, D. D. and C. Beringer. 1960. *Costs and Benefits of Seeding Range Lands in Southern Idaho.* Idaho Agricultural Experiment Station Bulletin No. 326. 31 pp.

Caton, D. D., C. O. McCorkle, and M. L. Upchurch. 1960. Economics of improvement of western grazing land. *J. Range Manage.* 13:143-151.

Child, R. D. and G. R. Evans. 1976. *Computer Optimization Planning System.* Range Science Series Report No. 19, Colorado State University. 73 pp. (Mimeo.).

Cook, C. W., G. Taylor, and E. T. Bartlett. 1980. Impacts of federal range forage on ranches and regional economies of Colorado. Colorado State University Experiment Station Bulletin No. 576-S. 7 pp.

D'Aquino, S. A. 1974. A case study for optimal allocation of range resources. *J. Range Manage.* 27:228-233.

Dean, J. 1954. Measuring the productivity of capital. *Harvard Bus. Rev.* 32:120-130.

Gardner, B. D. 1961. *Costs and Returns from Sagebrush Range Improvement in Colorado.* Colorado Agricultural Experiment Station Bulletin No. 511-S. 18 pp.

Gardner B. D. 1963. The internal rate of return and decisions to improve the range. In *Economic Research in the Use and Development of Range Resources. Development and Evolution of Research in Range Use and Development.* Western Agricultural Economics Research Council Report No. 5, Laramie, WY. Pp. 87-109.

Gray, J. R. 1968. *Ranch Economics.* Iowa State Univ. Press, Ames. 534 pp.

Gushee, C. H. 1968. *Financial Compound Interest and Annuity Tables.* Financial Publishing Co., Boston, MA. 884 pp.

Hanke, S. H., P. H. Carver, and P. Bugg. 1975. Project evaluation during inflation. *Water Resources Research* 11:511–514.

Hewlett, D. B. and J. P. Workman. 1978. An economic analysis of retention of yearlings on range and potential effects on beef production. *J. Range Manage.* 31:125–128.

Hunter, D. H., E. T. Bartlett, and D. A. Jameson. 1978. Optimum forage allocation through chance-constrained programming. *Ecological Modelling: International Ecol. Modelling* 2:91–99.

Kim, Chaiho. 1971. *Introduction to Linear Programming*. Holt, Rinehart, and Winston, New York. 556 pp.

Leistritz, L. F. and N. J. Qualey. 1975. Economics of range management alternatives in Southwestern North Dakota. *J. Range Manage.* 28:349–352.

Mishan, E. J. 1976. *Cost-Benefit Analysis*. Praeger, New York. 454 pp.

National Research Council. 1970. Nutrient requirements of beef cattle No. 4. *Beef Cattle*. 4th rev. ed. National Academy of Sciences, Washington, DC.

Nielsen, D. B. 1967. *Economics of Range Improvements—A Rancher's Guide to Economic Decision Making*. Utah Agricultural Experiment Station Bulletin No. 466. 48 pp.

Nielsen, D. B. and S. D. Hinckley. 1975. *Economic and Environmental Impacts of Sagebrush Control on Utah's Rangelands*. Utah Agricultural Experiment Station Research Report No. 25. 27 pp.

Overton, W. S. and L. M. Hunt. 1974. A view of current forest policy, with questions regarding the future state of forests and criteria of management. *Transactions of 39th North American Wildlife and Natural Resources Conference*. Wildlife Management Institute, Washington, DC.

Pechanec, J. F., G. Stewart, and J. P. Blaisdell. 1954. *Sagebrush Burning—Good and Bad*. USDA Farmer's Bulletin 1948 (revised). Pp. 33–35.

Renner, F. G. and B. W. Allred. 1962. Classifying Rangeland for Conservation Planning. USDA SCS Agricultural Handbook No. 235. 48 pp.

Sharp, W. W. and C. C. Bokin. 1967, A dynamic programming model for evaluating investments in mesquite control and alternative beef cattle systems. Texas Agricultural Experiment Station Technical Monograph No. 4. 38 pp.

Solomon, E. 1956. The arithmetic of capital budgeting decisions. *J. Bus.* 24:124–129.

Stoddart, L. A., A. D. Smith, and T. W. Box. 1975. *Range Management*. 3rd ed. McGraw-Hill, New York. 532 pp.

Subcommittee on Range Research Methods of the Agricultural Board. 1962. Economic research in range management. In *Basic Problems and Techniques in Range Research*. National Academy of Sciences Publication No. 890. P. 206.

Torell, L. A., G. F. Speth, and C. T. K. Ching. 1982. Effect of calf crop on net income of a Nevada range cattle operation. *J. Range Manage.* 35:519–521.

U.S. Department of Agriculture. 1980. *Farm Real Estate Market Developments*. USDA Economics, Statistics, and Cooperatives Service Report No. CD-85, August, Washington, DC. 46 pp.

Vallentine, J. F. 1971. *Range Development and Improvements*. Brigham Young Univ. Press, Provo, Utah. 516 pp.

Whitson, R. E., R. K. Heitschmidt, M. M. Kothmann, and G. K. Landgren. 1982. The impact of grazing systems on the magnitude and stability of ranch income in the rolling plains of Texas. *J. Range Manage.* 35:526–532.

Workman, J. P. 1976. Economic evaluation of prescribed burning projects. *Use of*

Prescribed Burning in Western Woodland and Range Ecosystems—A Symposium. March 18, 1976, Utah State University, Logan, Utah. Pp. 17–24.

Workman, J. P. 1981. Disagreement among investment criteria—A solution to the problem. *J. Range Manage.* 34:317–324.

Workman, J. P. and J. F. Hooper. 1971. Cost-size relationships of Utah cattle ranches. *J. Range Manage.* 24:462–465.

Workman, J. P. and D. W. MacPherson. 1973. Calculating yearlong carrying capacity—An algebraic approach. *J. Range Manage.* 26:224–227.

TEXT NOTES

1 Since real returns are being used in this example, a real discount rate should also be used. The $0.50 annual net return per acre is based on a future projection of the current (constant dollar or real) price of forage.

Consistency requires that since the effects of inflation are not included in future income, neither should inflation be included in the interest rate. Alternatively, the analysis could be based on nominal (inflated) returns and a nominal discount rate (current market rate that includes expected future inflation). This second approach would require a prediction of the future nominal price of forage throughout the 15-year life of the prescribed burning project. Either approach is correct, provided the consistency rule is observed.

Many analyses in resource economics have incorrectly used a nominal interest rate to discount a real net income stream. There are probably several reasons why this inconsistency is widespread. First, the conservative analyst may be tempted to use a nominal interest rate simply because a nominal rate is paid for borrowed capital and combining a nominal rate with real returns provides a margin for error. However, nominal returns are what the investor actually receives from the project, not real returns. Second, neither the future real interest rate nor future nominal returns are known at the time a project is initiated; both are known at the end of the project life. At project initiation we know only the nominal interest rate (the current market rate, which includes expected future inflation) and the real return (the current product price applied in constant dollars over the life of the project). Since analysts are understandably reluctant to predict either the real interest rate or nominal returns, they often incorrectly use what is known: the nominal interest rate and real returns. Criticism of this inconsistency in economic evaluation of federal projects (Overton and Hunt, 1794; Hanke et al., 1975) is detailed in Appendix I of Chapter 10.

2 In our hypothetical example no annual maintenance costs are associated with the prescribed burn. If the project did involve annual maintenance and operating costs, these would be discounted and added to the initial investment to form present value of costs (C) and gross annual benefits, rather than net annual benefits (R), would be discounted to form present value of benefits (B) (Barkley and Seckler, 1972).

3 Thus far all income or cost flows discussed have been examples of the special case of annuities. *Annuities* are equal annual increments of income received (or payments dispersed) over a period of years. Of course not all range improvements produce these uniform flows of annual income. The more complicated method required to calculate IRR for uneven flows is discussed in Appendix II of this chapter.

4   Modern handheld electronic calculators such as the Hewlett-Packard 22 or Texas Instrument Money Manager are programmed to automatically generate IRR for uniform annual income flows. According to such a calculator, the exact IRR for our hypothetical project is 9.13 percent.
5   Selection of an optimum combination of projects from among a variety of possible range improvements is discussed in Appendix I of this chapter.
6   For the purpose of estimating forage production, range condition can be thought of as the existing vegetation on a particular range site expressed as a percentage of potential vegetation for the site. Range condition is often expressed in broad categories such as excellent, good, fair, and poor.
7   It should be recognized that this last calculation assumes that the usable forage will be located at the same point in space and time as the grazing livestock. Many factors influence actual utilization of potentially usable forage including distance to stock, water, terrain, fencing, type of livestock, season of grazing, competitive (or other) interactions with wild herbivores, etc. A simple conservative rule might be: If livestock cannot, will not, or might not graze it, then do not count it.
8   Dry matter requirement varies with environmental conditions, size, age, and physiological function of the animal, and with nutritive quality of forage being grazed. The National Research Council provides estimates of dry matter requirements for a wide variety of livestock production situations.
9   This is a hypothetical example presented only for the purpose of demonstrating the evaluation technique. The numbers used here should not be interpreted as a projection of expected returns from prescribed burning.
10 For the sake of simplicity in demonstrating forage balance calculations, the feed supplies shown in Table 9.1 are slightly different than those listed above for the 300-cow ranch of Chapter 2. If the feed sources were identical, the U.S. Forest Service summer range permits and the winter and spring grazing provided by Bureau of Land Management (BLM) range would be classified as "inflexible feeds" along with private spring pasture, summer range, and crop aftermath discussed below.
11 Due to inconsistencies when applied to feeds of differing nutritional quality, the TDN measurement is being phased out of livestock nutrition literature in favor of evaluation systems based on energy content of feeds (Stoddard et al., 1975). However, as with TDN values, the practicing range manager must depend upon published averages of energy content of feeds in order to apply an energy accounting approach. Work by Kearl (1970) has shown that TDN and energy content give similar results for range management planning purposes.
12 Nursing calves under four months of age are commonly included with their dams in measuring AU since independently they exert little pressure on the forage resource.
13 This allocation of the flexible sources of feed (barley and hay) is arbitrary and simply spreads these feeds evenly over the six months November–April and maintains a uniform monthly surplus during this period of 18 AUM.
14 The "average month" method is one such approach (Workman and MacPherson, 1973). This method first divides total available AUM by 12, giving the number of AUM available during the average month: (4675/12) AUM = 390 AUM. Next this average monthly feed supply is divided by the average monthly AUM requirement per breeding cow, based on some rule of thumb such as 1.2 AUM of feed are required monthly for each cow in the breeding herd (American Institute of

Real Estate Appraisers, 1972). Thus estimated breeding herd carrying capacity for the subject ranch is (390 AUM)/(1.2 AUM/cow) = 325 cows. As in this example, the average month method often overestimates breeding herd carrying capacity since constraints imposed by limiting months are not taken into account.

15 Any improvements in livestock production, in addition to the expanded cow herd, that are expected to result from the prescribed burn should also be included in the partial budgeting procedure that follows. The improved nutritional status of cows during the spring months might substantially increase both conception rates and milk production, resulting in a higher calf crop percentage and increased calf weaning weights. Increased calf crop percentage is especially important. Gray (1968) reported that due to the fixed character of many cattle production costs, a one percent increase in calving percentage would raise net ranch income by almost three percent.

16 Required investment in the expanded cattle herd can be handled as part of the initial cost of the burning project or as a reduction in net ranch income due to increased heifer retention during the first year of the productive life of the project. The latter approach results in a nonuniform annual flow of net income (a problem discussed in some detail in Appendix II of this chapter). The first approach is much simpler and will be used in our present example.

17 Compare the total feed requirement column for 300 cows in Table 9.4 with the corresponding column for 320 cows in Table 9.5.

18 It should also be recognized that accounting for the required investment in the expanded cow herd in this manner causes depreciation costs to be overestimated. At the end of the 15-year project life, the expanded cow herd, unlike the burned area, will still be intact due to retention of replacement heifers. The cost of retaining heifers for required herd replacement has already been taken into account as a reduction in the added annual cash returns of Table 9.6.

19 The IRR calculated by a Hewlett-Packard 22 financial calculator is 29.37 percent.

20 This IRR is considerably less than the 29 percent real rate calculated previously for the same prescribed burning when increased forage production was valued at the private lease rate. If increased calf crop percentage and improved weaning weights had been included as expected benefits (as they are by ranchers bidding for additional AUM in the private lease market), the IRR values calculated by the two methods would undoubtedly have been more similar. Even so, it is not uncommon for analyses based on private lease rates to yield higher net returns than those based on increased livestock production.

APPENDIX I NOTES

1 IRR values can be approximated using Present Worth of One Per Period or other appropriate financial tables (Gushee, 1968), but the detailed values in Table 9.2 require the use of financial calculators such as the Hewlett-Packard 22 or Texas Instrument Money Manager.

2 As shown in Tables 9.15 and 9.16, when correctly calculated, the project selection of all three investment criteria are dependent upon the discount rate. Thus, in this sense all three criteria are inconsistent.

3 Alternatively, his investment goal could be defined in terms of maximum future net worth (Mishan, 1976). In Table 9.14 he would prefer the future net worth earned by

the stock water project ($18,328 minus the compounded value of the original $2400 investment, $16,146, or a future net worth of $2182) to that earned by the spraying project ($17,552 − $16,146 = $1406).

4 It should be noted that since normalization of PNW involves compounding forward and discounting backward at the same interest rate (the opportunity cost), PNW is not changed by normalization. For this reason the PNW values in Table 9.14 are identical to those in Table 9.12.

5 As calculated in the usual manner in Table 9.12 and 9.13, the IRR criterion assumes that annual net returns are reinvested at the calculated IRR rather than the true alternative rate of return, the opportunity cost. For the $1000 sagebrush spraying, the 12.96 percent return is based on the assumption that each $142 of annual return is reinvested at 12.96 percent compounded annually over the 20-year project. This assumption may be demonstrated simply by compounding the $142 annual return forward at 12.96 percent to give a future value of $11,441 and then calculating the annual compound interest rate required to allow a $1000 investment to grow to $11,441 over 20 years. The calculated IRR is 12.96 percent, a result that cannot be achieved if the $142 annual return is compounded forward at any interest rate other than 12.96. If, for example, the annual return were compounded forward at 10 percent, the calculated IRR would be 11.05 percent. It should also be pointed out that the IRR values calculated in this way are not strictly internal, since they depend on the interest rate used to convert annual net returns to future values.

6 The only exception to the general statement that PNW is unaffected by normalization is the special case where borrowing rate exceeds opportunity cost rate. In this case normalization of PNW values would involve first compounding forward at the lower opportunity cost rate and then discounting backward at the higher borrowing rate, resulting in smaller PNW values.

APPENDIX II NOTES

1 As explained in Appendix I, the common method of calculating IRR is based on the assumption that net project returns are immediately reinvested at the calculated IRR. The discussion that follows applies only to the case where this somewhat unrealistic assumption is retained. When IRR calculations are based on returns data corrected for the actual reinvestment rate (opportunity cost), the iterative procedure described below is not necessary.

2 Some models of advanced financial calculators manufactured by Hewlett-Packard automatically calculate IRR values for nonuniform flows.

Chapter 10

Benefit–Cost Analysis on Public Ranges

ROLE OF ECONOMIC ANALYSIS ON PUBLIC LANDS

Economics deals with problems of efficiency and equity. As previously emphasized, efficiency involves allocation of scarce resources among competing uses and time periods to produce the greatest quantity of net product from a given amount of resources. Equity concerns the distribution of products among competing consumers and resource owners. Resource allocation decisions on public lands (those owned by federal or state governments) affect both efficiency and equity, each of which contains elements of *objective* economics (what is) and *subjective* economics (what ought to be) (Gardner, 1981; Bromley, 1981).

It is a national ethos in the United States that efficient production is a desirable goal. Reaching this goal requires optimum resource allocation, but federal land management equity goals often lead to inefficient resource allocation decisions. An example is the eligibility requirements for holding federal grazing permits. The requirements were originally adopted to achieve equity goals, but their retention restricts permit transfer and precludes competition among stockmen (Gardner, 1981).

Some legislative mandates, such as those for multiple use, are not only ambiguous as management criteria, but also create barriers to efficiency and equity (Gardner, 1981). One equity problem is that some users of public rangeland (stockmen holding grazing permits) pay fees, while others (recreationists and watershed beneficiaries) do not.

Analysis versus Justification

Decisions concerning the services to be provided by government are made primarily in the political rather than the economic arena (Barkley and Seckler, 1972). This statement applies as well to specialized grazing systems and vegetation conversion projects on public rangeland as to government investment in highways, flood control, or national defense. Still, Congress has traditionally made it clear that economic feasibility is one of several criteria to be used in federal investment decisions. In land management, this philosophy probably first appeared in the Flood Control Act of 1936, which stated that water development projects would be authorized only if benefits "to whomsoever they accrue" exceeded estimated costs. This general guideline prevailed for some 40 years in major published federal guides including the "Green Book" (U.S. Interagency Committee on Water Resources, 1950), Circular A-47 (Bureau of the Budget, 1956), U.S. Senate Document No. 97 (1962), National Environmental Policy Act of 1969 (U.S. Congress, 1970), and Proposed Principles and Standards for Planning Water and Related Land Resources (Water Resources Council, 1971). More recent examples include Public Lands and Resources: Planning, Programming, and Budgeting (BLM, 1979a), BLM Instruction Memorandum No. 80-57 (BLM, 1979b), National Forest System Land and Resources Management Planning (U.S. Forest Service, 1979), and Economic Analysis for Forest Planning (U.S. Forest Service, 1980b).

While federal government decision makers have consistently included at least a token amount of economics in their analysis of public projects, they have also retained the power to base the actual investment decision on other than economic ground (Barkley and Seckler 1972). This tendency is perhaps best demonstrated by the fact that, unlike investments in the private sector, public projects are normally not ranked and implemented according to descending economic efficiency. Instead, proposed federal projects are usually only subjected to a single economic test: Is the project economically feasible (is the benefit/cost ratio 1 to 1 or greater)? Thus a federal project promising a benefit/cost ratio of 10 to 1 is judged no more favorable than one expected to yield a benefit/cost ratio of 1 to 1 (Barkley and Seckler, 1972; Herfindahl and Kneese, 1974). In other words, although federal evaluation procedures justify the transfer of private funds into public projects on economic feasibility grounds, allocation of the scarce funds within the agency is not designed to maximize net economic return.

This policy is favored by the public land manager, of course, since it preserves his flexibility with respect to project selection. Convincing arguments could be developed for the gains in federal investment efficiency that would

BENEFIT–COST ANALYSIS ON PUBLIC RANGES

result if federal projects were implemented in order of their descending economic feasibility (as described in Appendix I of Chapter 9). Realistically, however, economic analysis will probably continue to be used by federal land managers more for justification of projects already chosen for a variety of valid, but noneconomic, reasons, rather than as a tool for project selection. Even so, the economic justification required for all federal projects is, by itself, sufficient reason for the public range manager to become more skilled in economic analysis.

The Ravages of Time

My initiation into federal agency economics came some 15 years ago in the form of a research contract jointly sponsored by the Bureau of Land Management (BLM) and Bureau of Reclamation (BOR). The research assignment was to "conduct an economic analysis" of soil erosion control treatments (contour furrows and gully plugs) being applied in the Cisco area of eastern Utah. Being completely unfamiliar with "government economics," I had the mistaken notion that economic analysis of the erosion control treatments meant the same thing to the Department of Interior as it did to me: to determine whether or not the treatments were economically feasible (i.e., whether or not benefits exceeded costs). Two facts quickly became apparent. First, BLM and BOR wanted an economic justification, not an economic analysis (not surprising, since over 5000 acres of contour furrows and more than 10,000 gully plugs were already in place). Second, no matter how generous the assumptions concerning such items as control costs, project life, and erosion reduction, there was no way to provide an honest economic justification of these erosion treatments.

Glen Canyon Dam was the primary target to be protected from erosion in the upper Colorado River basin. Annual net benefits provided by the dam were estimated at $48,700,000 and BOR engineers set the expected life of the dam at 200 years under prevaling erosion rates. Thus under the best circumstances, the value of complete protection of the dam from erosion was as follows: ($48,700,000)/0.07 − $48,700,000(14.283) = $132,186.[1] (The present value of the net benefits to be produced during the 200 years of dam life expected without erosion control were subtracted from the capitalized value of receiving the net benefits forever, giving the present value of extending dam life from 200 years to infinity.)

After a presentation of these results to the American Society of Civil Engineers (Workman and Keith, 1975), a doubting (and pro-erosion control) engineer in the audience asked two questions:

> Do you mean to tell me that you economists think you can calculate what it's worth to extend a $49,000,000 annual flow from 200 years to infinity? And that when you tried it came out to be only $132,000?

Interpreting the calculations in terms of depositing $132,186 in the bank today at seven percent interest, waiting 200 years, and then withdrawing $48,700,000 per year forever without ever depleting the principal did little to satisfy the questioner or the other engineers in the audience.

This example illustrates the drastic effects of compound interest rates used in standard benefit–cost analysis to calculate the present value of future benefits. The exponential nature of such discounting calculations causes benefits in the distant future to be almost worthless today. This apparent harsh treatment of the future has led at least one anonymous observer to state that "on a strictly benefit–cost basis, the world isn't worth saving." Still most of us reveal by our actions (despite what we say) that we do have a strong preference for present benefits over those to be received in the future. Rare is the family that does not pay interest on borrowed funds rather than waiting until the amount required to purchase a new car has been saved. Even natural resource students, while idealistically expressing what seems to be a sincere concern for the "watering down of posterity's vote" caused by discounting calculations, simultaneously reveal a striking preference for present personal benefits over those to be received in the future. As one example, consider those students who take out federally guaranteed education loans and then use the money to buy season ski passes.

It seems, then, that despite the severe effects of discounting calculations on resource management policy relating to energy reserve exploitation, wilderness area designation, and timber harvest rotations, time preference is an almost universally held philosophy as revealed by our actions (if not our statements) regarding present versus future consumption. For none other than this very good reason, discounting will likely remain a part of public investment benefit–cost analysis. The implications in range management are obvious, since range improvement programs are usually designed to produce streams of benefits that will reach their full potential only at some point in the future.

ECONOMIC FEASIBILITY AND PUBLIC RANGE INVESTMENT

The abovementioned congressional directive that economic feasibility is only one of several criteria to be used in decisions concerning implementation of federal projects was reenforced by the National Environmental Policy Act (NEPA) of 1969 (U.S. Congress, 1970). NEPA states, in part,

> ...insure that presently unquantified environmental amenities and values be given appropriate consideration in decisions along with economic and technical considerations.

The recent emphasis on environmental considerations has prompted a British observer (Mishan, 1976) to coin the phrase "horse and rabbit stew." In his opinion, no matter how carefully the economic analysis (the rabbit) is prepared in public investment decision documents, the overall project analysis (the stew) will still taste like the environmental considerations (the horse).

The Water Resources Council (WRC) is composed of cabinet secretaries from the executive branch along with the Assistant Director of the Office of Management and Budget. A primary responsibility of WRC has been to standardize the evaluation practices used for planning water resource development. In complying with NEPA, WRC (1971) published their Proposed Standards in the *Federal Register*. A system of four "accounts" (national economic development, environmental quality, regional development, and social factors) was included to accommodate NEPA requirements and to cover the primary and secondary (induced) cost and benefit categories of U.S. Senate Document 97 (1962) that the Proposed Standards replaced.

An Idealized Benefit–Cost Analysis

As required by NEPA and as recommended by WRC (1971), the cost and benefit categories for an idealized benefit–cost analysis for a hypothetical range improvement project (a pinyon-juniper chaining) are shown in Table 10.1. *Quantifiable costs* are those that can be expressed in numerical units. *Market priced* quantifiable costs involve items that are actually bought and to which monetary values can be assigned. Included are pretreatment cost (direct costs of personnel time spent in project planning and evaluation, specialist consulting fees, telephone, and travel), treatment costs (contractual equipment and labor, fencing, water facilities, and personnel time spent in supervising and implementation), posttreatment costs (grazing deferment to allow seedling establishment of desired plants, follow-up control of tree seedlings, maintenance of fences and water facilities, and additional management and supervision related directly to the chaining project). A fourth category, foregone benefits, includes treatment-caused losses of benefits, such as the value of pinyon nuts normally harvested that are lost because of tree removal. Foregone benefits should be counted either as treatment costs or as benefits produced by untreated land. In the latter approach, the value of foregone pinyon nuts would be accounted for in the calculation of net project benefits (benefits produced after chaining minus benefits produced prior to treatment).

Nonmarket priced quantifiable costs refer to those benefits given up because of treatment that cannot be valued in dollar terms but that can be expressed in relevant quantitative units (WRC, 1971). Increased erosion or adverse changes in runoff because of tree removal would be included in this category. Loss of habitat for wildlife dependent upon pinyon-juniper stands is another example. Such impacts would be entered as costs, if detrimental, but as benefits if the project is expected to have positive effects. Thus, some of the categories of Table 10.1 might be listed as either costs or benefits, depending on expected results.

Nonquantifiable costs are those items lost due to the project that not only cannot be valued in dollar terms but cannot even be expressed in numbers. In

Table 10.1 Cost and Benefit Categories Required for an Ideal Benefit–Cost Analysis of Pinyon-Juniper Chaining

I. Costs
   A. Quantifiable
       1. Market priced
           a. Pretreatment
           b. Treatment
           c. Posttreatment
           d. Foregone benefits
       2. Nonmarket priced
           a. Physical
           b. Biological
   B. Nonquantifiable
       1. Aesthetics
       2. Recreation
       3. Archeologic

II. Benefits
   A. Quantifiable
       1. Market priced
           a. Livestock forage
           b. Woodland products
       2. Nonmarket priced
           a. Biological
           b. Physical
   B. Nonquantifiable
       1. Aesthetics
       2. Recreation

Source: Workman and Kienast (1975).

this example, these include the detrimental effects of chaining on aesthetics, recreation opportunities, and archeological sites. This group of costs also appears as benefits in Table 10.1. The subjective nature of what is perceived as aesthetically pleasing, what represents a recreation opportunity, and what might be society's best use of a potential archeological site makes it necessary to allow for both possibilities. While displeasing to some people, a well planned conversion of pinyon and juniper trees to a mixed stand of grass and shrubs is viewed by others as a beneficial change, since it adds variety to what they regard as a monotonous landscape. It is also not difficult to imagine an honest argument between persons sharing an equal concern for a potential archeological site. One faction might maintain that, because of possible damage, chaining should not be done, while another might counter that the chaining might uncover artifacts that otherwise would never be found.

Quantifiable benefits are those that can be expressed in numerical units and consist of both market priced and nonmarket priced positive effects of the project. *Market priced* benefits are (or could be) sold and can be valued in dollars. Increased livestock forage, valued either directly at its market price or indirectly in terms of expected increases in livestock production, is one example. Increased production of woodland products, including fence posts, firewood, pinyon nuts, Christmas trees, and charcoal, are also market priced benefits. As previously mentioned, if chaining were expected to cause a loss of woodland products, these should be counted as market priced costs. Thus, while chaining might provide an opportunity for improved future Christmas tree harvest (a market priced benefit) through stimulation of seedling and sapling growth, eradication of mature trees would simultaneously diminish opportunities to harvest pinyon nuts (a market priced cost).

Nonmarket priced benefits are positive project effects that cannot be valued in dollar terms, but which can be expressed in quantitative units. These benefits may be either biological (increased wildlife numbers or greater species diversity due to habitat improvement) or physical (improved water quality, yield, or seasonal flow distribution). In some cases these benefits can be legitimately grouped with market priced benefits. For example, improvements in deer habitat resulting in increased hunting opportunities similar to those actually commanding hunting lease fees on adjacent private land could be logically valued in dollar terms. However, many project benefits cannot be assigned monetary values even though they are recognized as important by both the public land manager and society at large. Many other benefits cannot be expressed in any sort of numerical units. These are the *nonquantifiable benefits* of Table 10.1 that include such things as improved quality of recreation opportunities and more aesthetically pleasing landscapes. These impacts of range improvements are perhaps best measured by some method of ascertaining public opinion as has been tried for pinyon-juniper chaining (Workman and Kienast, 1975).

Value Comparison Problems

Numerous researchers have struggled with the problem of pricing nonmarket resources and products for over 30 years (Hotelling, 1949; Clawson, 1959). While considerable progress has been made in valuation of a wide variety of recreation benefits including fishing (Brown et al., 1964), deer hunting (Wennergren, 1964; Garrett et al., 1970), and boating (Wennergren, 1965), the monetary values yielded by even the most sophisticated evaluation techniques are not truly comparable with prices set in the market place (Dwyer et al., 1977; Schuster, 1983; Godfrey, 1983).

Even if information for each of the idealized benefit–cost analysis categories of Table 10.1 were available to the public land manager, how would the decision-making process proceed from there? What weights should be assigned to the various cost and benefit categories in deciding whether or not to implement the project? The comparisons in Table 10.2 may help emphasize that even if all such information were available, the public land manager may still have difficulty in making (and justifying) a decision.

Situation 1 in Table 10.2 represents the desirable problem where the range improvements project is not only economically feasible, but all other resources and products are benefitted as well. From any viewpoint, the project should be

Table 10.2 Comparison of Hypothetical Economic Impacts of Range Improvement

Situation	Net market priced (economic) quantifiable impacts	Net nonmarket priced biological (wildlife) and physical (watershed) quantifiable impacts	Net nonquantifiable (recreation and aesthetics) impacts
1	+$1	+1 unit	Positive
2	−$1	−1 unit	Negative
3	+$1	−1 unit	Negative
4	−$1	+1 unit	Positive

Source: Workman and Kienast (1975).

implemented. In situation 2, the opposite is true. Not only does the project produce an economic loss, but it is detrimental to all other resource values. The unanimous decision would be to not implement the project.

Seldom, if ever, are management decisions this simple. Situation 3 poses the more realistic problem where the improvement project promises positive economic returns but imposes detrimental impacts on other important resource values. Here the manager must weigh monetary returns against the loss of other benefits to which his public may assign equal (or greater) values. Situation 4 represents the opposite case. The project promises a net economic loss but produces beneficial impacts on all other resources and products. Again the manager must weigh the noneconomic needs of an interested public against the economic loss produced by the project.

Thus some of the public range manager's most difficult decision-making problems, as well as the opportunity to retain flexibility in directing resource programs, are due to NEPA provisions and resulting recommendations of the WRC (1971). According to NEPA, benefit–cost ratios greater than 1:1 may not, by themselves, be sufficient justification for range improvement projects. Equally important, benefit–cost ratios of less than 1:1 are not necessarily sufficient grounds for project rejection if nonmarket resources and products would be improved by the project. The next section discusses two project evaluation approaches that attempt to either avoid or deal with the extremely difficult task of pricing nonmarket resources while still complying with the intent of NEPA.

COST EFFECTIVENESS AND NONMARKET PRICING

Cost Effectiveness: A "Specified Bang for the Smallest Buck"

Suppose a large public land grazing allotment makes an important forage contribution to the local livestock industry. However, range condition is presently only fair and the trend is downward. Suppose further, that the allotment is extremely visible, lying next to a well traveled recreation access road in the upper reaches of a watershed. The area provides important elk summer range along with culinary water for a nearby town. There is a tremendous incentive, of course, for the manager to immediately implement a range improvement program in order to avoid even more serious resource problems in the future. Suppose that in the manager's judgment the most efficient means of accomplishing an improvement in range condition appears to be through the establishment of a specialized grazing system.

To carry the scenario a bit further, suppose that the grazing system prescribed does improve range condition and produces a host of benefits: improved soil stability, improved water quality, improved elk habitat, and, of course, more livestock forage. Of all the important benefits expected from the project, livestock forage alone is actually sold in the marketplace and it alone carries an accurate dollar value. Livestock forage might be priced in one of several ways. Increased forage could be valued as: net annual return = [(AUM after − AUM before) × $2 federal grazing fee] − (annual costs of maintenance and management). However, the very reason the project is being implemented is to avoid further declines in range condition that would harm all resource values and undoubtedly lead to an allotment cut in season or animal numbers. Thus increased forage would be more accurately valued by a "with and without" approach: net annual return = [(AUM with the project − AUM without) × $2 federal grazing fee] − (annual costs of maintenance and management). But the $2 fee represents the amount *charged* for the increased federal forage, not its true market value. A more accurate evaluation, then, would involve the average lease fee paid for private range: net annual return = [(AUM with − AUM without) × $8 private grazing fee] − (annual costs of maintenance and management). Despite extensive literature covering methods of evaluating increased grazing for benefit–cost purposes (Dickerman and Martin, 1967; Martin and Jeffries, 1966; Regional Research Project W-79, 1968; Smith and Martin, 1972; O'Connell and Boster, 1974), the argument is still heard occasionally that increased forage production from federal range improvement should be valued at the federal grazing fees actually charged rather than at market value. But federal fees are administered prices rather than market prices and do not reflect the true value that ranchers (or society) place on the increased forage. The argument that federal forage value equals federal grazing fees charged is like saying that since recreation benefits are provided free, they should be assigned a value of zero in benefit–cost analysis. The fact that recreationists are not charged (or pay only nominal fees) for recreation does not mean that these opportunities have zero or small values. The same is true of forage on federal lands.

Let us suppose that in this hypothetical example, the grazing system proves to be an economic success even though the benefit–cost analysis assigns monetary values only to increased forage. But the question might be asked whether it is equitable (or even logical) to require increased livestock forage to justify the full cost of the project? The obvious answer is no, it is not reasonable to require increased forage to subsidize the various nonmarket benefits. However, forage is the only market priced product and if the project can be justified economically by forage increases alone, the other benefits may be viewed as a bonus.

But what if the net returns from forage production are insufficient to provide this luxury? What possible justification could be offered for the prescribed grazing system if the dollar value of increased forage fell short of the project cost? These questions are normally answered in one of two ways. The first (and less common) approach is a branch of benefit–cost analysis called *cost effectiveness* analysis[2] (Howe, 1971). Instead of searching for the "biggest bang for the buck" (as described in Chapter 9), the goal of cost effectiveness analysis is to produce a specified "bang" for the "smallest buck." For obvious reasons this approach has long been used by military analysts. Applied to our hypothetical planned grazing system, the cost effectiveness approach would consist of first pointing out the positive relationship believed to exist between improved range condition and improved soil stability, water quality, and wildlife habitat (Council for Agricultural Science and Technology, 1974). Next it would be made clear that, since the allotment is in only fair condition (and even more important, trend is downward), some action must be taken immediately if irreversible resource deterioration is to be avoided. The specialized grazing system promises to reverse the downward trend. Grazing fees received from the increased forage production will effectively subsidize production of the various nonmarket benefits and the more that AUM production is increased, the lower the net costs of improving range condition. Thus the grazing system appears to be the least cost means of accomplishing a priority objective.[3] A plea has been made within BLM (Fulcher, 1973) that this approach be adopted for BLM range improvement decisions.[4]

While there are undoubtedly many worthwhile projects that could be honestly analyzed using the cost effectiveness approach (for example, protection of an endangered species as directed by Congress), use of this approach could obviously be abused. If the only economic test that proposed federal projects had to pass was to be the least-cost means of accomplishing a particular goal, any proposal could be justified. Thus cost effectiveness offers little help in answering the all important practical question of how to best allocate limited budgets and personnel among competing needs.

Nonmarket Pricing: A Search for Values

The second approach, searching for values for nonmarket benefits, is commonly used to justify range improvement projects that yield net losses in terms of market priced forage benefits. If enough nonmarket benefits are included, almost any biologically successful project will appear to yield a positive economic return. This approach is widely used by federal agencies, apparently due to their belief that a funding request fares better under Office of Management and Budget (OMB) scrutiny if estimated dollar benefits exceed dollar costs. Thus there is great incentive for the manager to value at least enough nonmarket benefits to justify project costs, even for those projects that the agency must implement to accomplish congressional directives, such as preserving an endangered species. In its crudest form, the search for values approach, as applied to the grazing allotment example, would simply involve claiming that the value of the various nonmarket benefits was sufficient to offset the project's dollar loss.

The search for values approach often causes federal managers to make serious errors in nonmarketing pricing logic that can, unfortunately, cause doubts concerning their credibility in what otherwise may be sound benefit–cost analyses. Examples of such errors in logic are not rare. The U.S. Forest Service has argued that the value of a tree that contains an eagle's nest is equal to the value of the timber that might have been harvested, but was not, in order to protect the nest. Clearly, the value described here is actually an opportunity cost that society must pay to save the tree and in no way measures the tree's value.[5] Even worse, BLM (1976) has stated "where bighorn sheep could be hunted, but hunting is not allowed, they are worth at least their foregone value in hunting...." In confusing hunting value with the opportunity cost of not hunting, the BLM authors have apparently misinterpreted the sound logic underlying the "willingness to pay" evaluation procedure. Briefly, willingness to pay estimates the value of, say, a hunter day, in terms of what a hunter would have been willing to pay for an opportunity enjoyed for free. This is an entirely different concept than the foregone value that BLM described as the amount that could have been charged for the opportunity, if it had been offered for use.

Numerous methods have been suggested for assigning monetary values to nonmarket resources. The debate continues concerning how far to go with quantification of nonmarket benefits and costs. However, it has become clear that there are two kinds of errors in the balancing equation: (1) A false sense of precision with numbers may give the impression that more is known than is really known; and (2) a false sense of imprecision without numbers may give the impression that less is known than is really known (Committee on Risk and Decision Making, 1982). Of the four pricing techniques to be described, only the "willingness to pay" approach yields logically consistent value estimates (Knetsch and Davis, 1974; Workman, 1975; Dyer, 1981).

Production Cost One unusual pricing method endorsed by the National Park Service (1950) is the production cost technique that consists of treating the cost of producing a desired product or service as its value. By this method, the value of a fish provided by a fish planting program would be calculated by dividing the total cost of the program by the total number of fish planted. The implied logic is that a rational administrator would not have undertaken the program if the value of the fish had not at least equalled their cost. This approach is clearly based on circular reasoning and should be discarded.

Cost of Alternatives Recreation opportunities on public land are often valued according to prices paid for similar opportunities on private lands. This approach is quite valid if the recreation opportunities available on public lands are similar to those actually being charged for on private lands. If privately owned campgrounds charge $5 per person per day, this same value may legitimately be assigned to similar public land opportunities. This approach is often not applicable, however, since public land recreation opportunities are often so unique that no private land substitutes exist. Two extreme examples are the Grand Canyon and Yellowstone National Park.

Gross Expenditure This approach treats the total expenditure for a recreation experience as the value of the experience. The value of a day of deer hunting furnished by a tract of public land, for example, is calculated as the hunter's total travel and on-site costs. The rationale is that the value of a day of deer hunting equals what the hunter paid for it.

Wennergren (1965) identified a serious inconsistency in the underlying logic of this method. Since value is based on total expenditure, the value of the recreation experience is highest for the participants who travel the greatest distances to reach the recreation site. Thus, the estimated value could be raised by simply increasing travel or on-site costs (by tearing out access roads, for example). Wennergren argued further that what was needed for correct evaluation was not the gross amount spent to use the resource but the net benefits derived from its use. In addition to the possibility of overestimates of value, it seems likely that this method might also underestimate the true value of a recreation experience. Consider the local deer hunter whose total costs of a day of hunting on public land consist of $5 for gasoline and perhaps another $2 for ammunition. If successful, the value of the meat alone may exceed $30. Meat value, however, represents only an estimate of the minimum worth of the total recreation experience (Knetsch and Davis, 1974). Even if the hunter does not shoot a deer and no meat value is involved, he would likely admit that $7 is a small price to pay for an entire day of activity. Whichever the direction of the pricing bias, there is no logical reason to believe that the gross expenditure technique yields accurate estimates of resource value.

Willingness to Pay The concept of consumer surplus (the most common willingness to pay approach) dates back to the mid-nineteenth century and was revived about 100 years later by Marshall (1947). Valuation of a product or service by this method is based on the idea that consumers, rather than going without a commodity, would pay more than they actually do pay and thus capture a surplus of value when making most purchases. The consumer surplus concept can be applied to any product or service, even those that do have established market prices. The consumer surplus associated with the purchase of peanut butter is shown in Figure 10.1. The willingness to pay of peanut butter consumers is represented by the demand curve, $D_1 D_2$. At a price of $3 per jar only 100 jars are purchased weekly. If the price drops to $2, weekly

BENEFIT-COST ANALYSIS ON PUBLIC RANGES

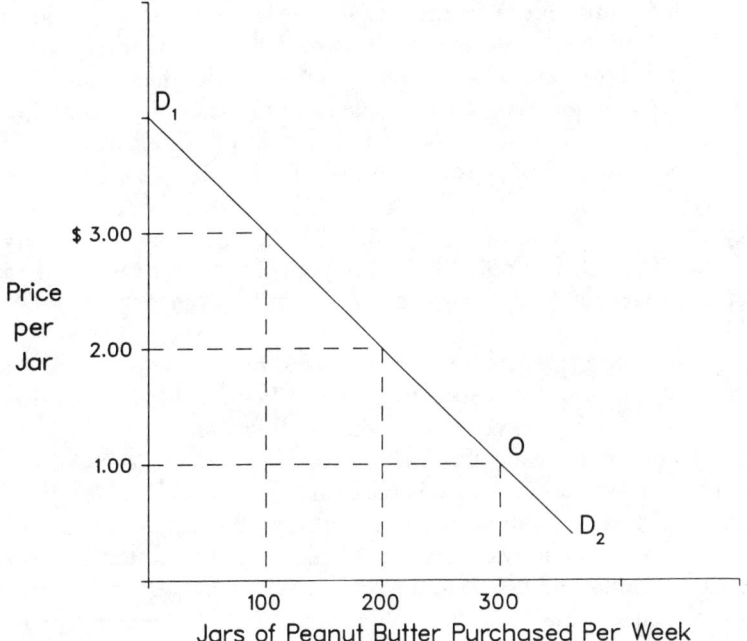

Figure 10.1 Demand curve for a market priced commodity (peanut butter).

purchases increase to 200 jars for two reasons. First, those consumers who were already buying peanut butter at $3 per jar now buy more. Second, consumers who refrained from buying peanut butter at the higher price now begin to substitute peanut butter for other foods. Note that although 200 jars are purchased at the $2 price, consumers would have bought 100 jars even if the price were still at $3 per jar. Therefore, the consumers who purchased 200 jars at $2 received a bonus or surplus value. They paid only $2 per jar for all 200 jars but as indicated by their willingness to pay, the first 100 jars were worth $3 per jar. Thus the consumers reaped a surplus value of $3 − $2 = $1 per jar on the first 100 jars purchased, for a total surplus of $100 at the $2 price.

Continuing our examination of the demand curve, if the price were $1 per jar, 300 jars would be purchased. At this price consumers would receive a surplus value of $3 − $1 = $2 on the first 100 jars and a surplus of $2 − $1 = $1 on the second 100 jars. If $1 were the lowest observed price and if we extended the logic of surplus value of all prices and quantities of purchase traced out by the demand curve, total consumer surplus is the area bounded by $D_1 - 0 - \$1$ on the graph.

Valuation of the peanut butter by the consumer surplus technique consists of treating the area $D_1 - 0 - \$1$ (what the consumers were willing to pay minus what they did pay) as the net value of the commodity. The market value of the 300 jars of peanut butter purchased at the lowest observed price of $1 per jar is $300, which, of course, does not equal the consumer surplus value.[6] It should be noted that for this reason, estimates of values of nonmarket products by the consumer surplus technique are not comparable with market priced values. Although based on much sounder economic logic than the other nonmarket pricing techniques mentioned, values estimated by the consumer surplus technique represent *a value* of the commodity in question rather than *the value* and they cannot be directly compared with market values in benefit–cost analyses.

Calculation of the consumer surplus value of a nonmarket commodity consists of six steps. The valuation of deer hunting opportunities provided by a particular tract of rangeland will be used as an example (Wennergren, 1964; Garrett, et al., 1970).

1. Zones of visitor origin. The communities served by the hunting area are divided into zones of visitor origin (Hotelling, 1949; Clawson, 1959). Zonation is usually based on distance or travel time to the hunting site.

2. Annual rate of use for each zone. Data concerning the amount of use of the area by the average hunter in each particular zone are obtained from questionnaires mailed to a random sampling of hunters residing in the various zones. Use rate for each zone is calculated in terms of annual hunter days of use divided by zone population. Use rate for zone 1, say, is the annual number of hunter days of use on the area by residents of zone 1 divided by the population of zone 1.

3. Cost per hunter day for each zone. Total on-site and travel cost data for each zone are also obtained from the hunter questionnaires. Average cost per hunter day of use for each zone is calculated by dividing the total zone use cost by the number of hunter days from that zone.

4. Demand curve for the hunting area. Regression analysis of annual rate of use against cost per hunter day yields the demand curve in Figure 10.2. Zone 1, the zone nearest the hunting area, contributes the highest use rate (R_1), and of course, because of lower travel costs, incurs the lowest average use cost (C_1).

5. Consumer surplus for each zone. Calculation of total consumer surplus for each zone involves multiplying the rate of surplus of the given zone by its population. For zone 1, rate of surplus is the area under the demand curve between the highest observed average cost (C_3) and the average cost incurred by zone 1 (C_1).

6. Calculation of total annual consumer surplus. Consumer surplus values for all zones are summed. The result is the estimated annual value of the hunting opportunities provided by the rangeland area.

Capitalization of this annual consumer surplus value provides an estimate of the present value of the perpetual future flow of hunting benefits for the land area in question. A consumer surplus study by Wennergren (1967) estimated the annual value of the regular rifle deer hunt to Utah's resident hunters to be $2,308,020. By capitalizing this annual consumer surplus at six percent interest, the value of Utah's resident deer hunting resource in perpetuity was estimated at $38,467,000.

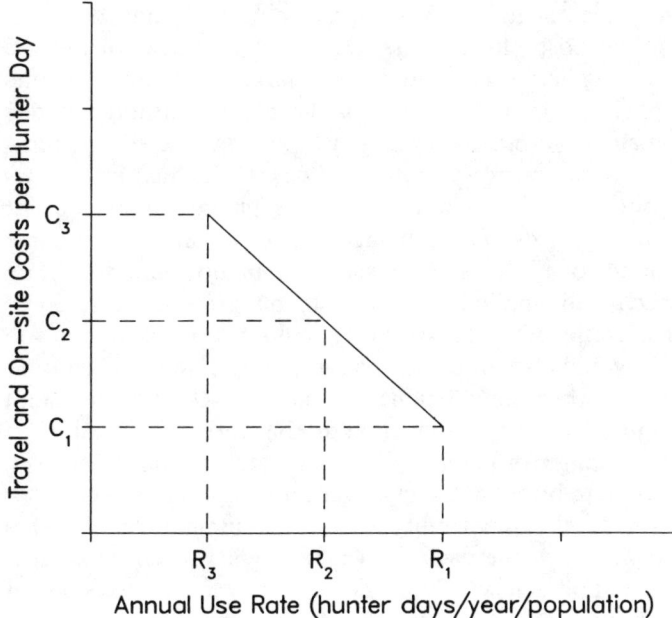

Figure 10.2 Demand curve for a hunting area serving three zones of hunter origin.

Garrett et al. (1970) conducted a similar consumer surplus evaluation of Nevada's deer hunting resource. The annual consumer surplus for each county was divided by the estimated deer population, the number of hunter days use, and the estimated number of deer AUM produced in the county. These calculations gave consumer surplus values per resident deer, per hunter day, and per deer AUM, respectively.

It should again be emphasized that the values for nonmarket resources estimated by the consumer surplus method cannot be compared directly with the market determined prices of range forage, irrigation water, and minerals. However, a near consensus exists in the literature that the willingness to pay procedure is the most appropriate conceptual framework available for valuation of nonmarket outputs (Dyer, 1981). Useful and reliable analytical techniques are available for valuation of many nonmarket wildlife and recreation outputs (Dyer, 1981; Howitt, 1981). Values for some nonmarket products (e.g., aesthetics, endangered species, and Indian burial grounds), however, can perhaps never be expressed in monetary terms. This does not mean that such resources are less important than those that can be priced.

Cost Effectiveness and Nonmarket Pricing—Summary

The main problem with both the cost effectiveness and the nonmarket pricing methods is that in current practice they are used more often to attempt to justify projects than as an aid in making decisions among alternatives (Workman and Fairfax, 1981). To avoid the trap of automatic justification of any and all proposed projects (and because it rarely provides any decision-making information), the cost effectiveness technique must be ruled out. Instead values must be explicitly assigned to nonmarket costs and benefits (they are already implicitly assigned, at least in the manager's mind) and a weighting system (albeit an arbitrary one) must be constructed to allow comparison of market and nonmarket benefits and costs. The final interpretation of any such scheme must be in terms of how high nonmarket benefit values (or how low nonmarket costs) must be in order for a particular resource decision to be the correct one. However, if such a valuation and weighting scheme is made explicit and applied uniformly to all projects or resource allocations under consideration, such an approach could be used as an input to making the decision. It would also allow the next logical step to be taken, that of sensitivity analysis, in which calculations are made of the amounts that nonmarket values would have to change before the correct allocation of resources would have to change (or before a feasible project became infeasible). Finally, such an approach, in bringing the manager's assumptions and expectations into the open, provides the opportunity for a doubting public (and OMB) to become acquainted with how the decision was reached. If such an explicit valuation and weighting approach were used in *making* resource decisions, it seems likely that *justification* of the process would take care of itself.

The Importance of the Interest Rate Used for Discounting

As discussed in Appendix I of Chapter 9, a forceful argument can be developed for using present net worth (PNW) rather than the benefit–cost ratio (B/C) as the investment criterion for choosing between two alternative projects. In the opinion of Barkley and Seckler (1972), one of the primary reasons that economic analysis in government has been based on the B/C criterion is its ease of interpretation by the public since "... anyone can easily understand that a ratio greater than unity means a positive return and a ratio less than unity yields losses... ." A convincing counterargument might be developed that a B/C ratio of, say, 1.2 is much more difficult for the public to interpret than either the amount of profit reported by PNW or the interest rate measured by internal rate of return (IRR). Since these latter criteria, rather than the B/C ratio, serve as the standards of both consumer and producer activities, the use of B/C for public investments may actually prevent the public from making comparisons of the economic efficiency of public investments with that of the private sector.

Whatever the investment criterion used, one of the principal determinants of the economic feasibility of range improvements is the interest rate used for discounting. When using the PNW criterion, the higher the discount rate, the lower the project profit or present value of net returns. Similarly, when using the IRR criterion, the higher the discount rate used for comparison with the IRR generated by a given project, the less likely the project will prove to be economically feasible. Even the outcome of the "go versus no go" B/C ratio test, to which federal projects are normally subjected, is dependent upon the interest rate used for discounting. University of Colorado Professor Kenneth Boulding summed up the crucial importance of the discount rate in one of his many poems (Smith and Castle, 1964):

> Around the mysteries of finance
> We must perform a ritual dance
> Because the long-term interest rate
> Determines any project's fate:
> At Two percent the case is clear,
> At Three, some sneaking doubts appear,
> At Four, it draws its final breath
> While Five percent is certain death.[7]

Although the interest rates mentioned sound low compared with current nominal rates, the message applies as well today as when first written.

The impact of the discount rate on project feasibility as measured by the B/C ratio criterion is demonstrated by the simple example of Table 10.3. Using a discount rate of five percent (the approximate rate recommend by the Water Resources Council prior to 1971), both projects pass the "go versus no go" test and since they yield identical B/C ratios of 1.2 : 1, the two projects are of equal economic efficiency as measured by B/C. But suppose the recommended interest rate is increased to seven percent, as it was in 1971 (Water Resources Council, 1971). At the higher discount rate, both the reseeding and the dam, while producing the same annual net returns for the same number of years as before, now yield much smaller B/C ratios. But the B/C ratio for the long-lived dam is reduced more severely and it is no longer a feasible project, while the B/C ratio for the reseeding still exceeds 1.0. This example demonstrates that an increased discount rate introduces a bias against

Table 10.3 The Effect of an Increase in Discount Rate on the Economic Feasibility of Two Hypothetical Public Land Projects

	Project	
	Reseeding	Dam
Expected life (years)	20	50
Initial investment ($)	50,000	50,000
Annual net return ($)	4,815	3,287
Present value of net return at 5% discount rate ($)	60,000	60,000
B/C at 5% discount rate	1.20	1.20
Present value of net return at 7% discount rate ($)	51,005	45,358
B/C at 7% discount rate	1.02	0.91

long-lived projects, causing them to be less attractive as compared with short-term investments. Environmentalist groups have occasionally used this bias to oppose construction of dams and other water development projects by campaigning for use of a higher discount rate for federal benefit–cost analysis. Discounting in general, and discounting at high rates in particular, then, is sometimes viewed as pro-conservation, since long-lived and sometimes irreversible federal projects are less apt to be imposed on future generations.

On the other hand, high discount rate, or discounting future benefits at all, is often perceived as being anti-conservation. If the discount rate used in private enterprise decisions is higher than the so-called social discount rate reflecting society's preference concerning *when* resources should be used, a bias is introduced against future generations. The higher the discount rate, the smaller the present value of future resource use and the greater the tendency to use stock resources such as fossil fuel reserves now rather than saving them for the future. For identical reasons, the higher the discount rate used for U.S. Forest Service planning, the shorter the optimum timber harvest rotation. Similarly, the higher the discount rate, the lower the present value of wilderness benefits from undeveloped federal lands and the more difficult the justification of preservation of national parks and wilderness areas.

Discount rate selection lies outside the discretion of the on-the-ground range manager. For example, the Bureau of Land Management (1976) specified the rate recommended by the Water Resources Council ($6\frac{1}{8}$ percent for fiscal year 1976) as the rate to be used for benefit–cost analysis of BLM range improvements. More recently, Row et al. (1981) recommended that the U.S. Forest Service adopt a discount rate of 4 percent. Their recommendation was based on the following discount rate criteria: (1) a marginal rate (based on returns on new private investments), (2) a rate that includes corporate income taxes, (3) a real (inflation-free) rate, and (4) a risk-free rate (any risk adjustments to be made by reducing expected yields or increasing expected costs, rather than adding a risk premium to the discount rate).

Because of the crucial role of the interest rate in determining the economic feasibility of public range improvements, discount rate selection is examined in more detail in Appendix I.

APPENDIX I. Selecting the "Correct" Discount Rate

As mentioned in the introduction to discounting in Chapter 8, the ideally "correct" interest rate used to discount for time combines the individual rates for (1) real (inflation-free) opportunity cost (2) risk, and (3) inflation. Each of these three important reasons for discounting benefits to be received in the future are discussed separately below.

Opportunity Cost versus Social Discount Rate

Traditionally, there have been two schools of thought concerning the discount rate to be used for benefit–cost analysis of public projects; the "Chicago School" and the "Harvard School" (Gardner and LeBaron, 1965). The basic belief of the Chicago School, so named because of its initiation and support by the economics faculty and students of the University of Chicago, is that public projects should stand the test of comparison with private investments (Howe, 1971). The Chicago School recommends that the opportunity cost of public investments be used as the discount rate for public projects. This opportunity cost rate is generally considered to best be measured by the pretax rate of return on the private investment foregone when funds are transferred to the public sector through federal taxes or federal borrowing to finance public investment. The rationale for use of the pretax private rate of return is that the federal government should not take funds from the private sector unless the federal return is at least equal to what would be earned on these same funds in the private sector. With a 50 percent marginal tax rate on corporate income, the private firm must earn 10 percent in order to pay its stockholders 5 percent in dividends. One dollar of public investment, then, displaces not only 5 cents in after-tax earnings but also causes 5 cents in tax revenue to be foregone. Thus while the government can borrow $1 for 5 percent interest by selling tax-free federal bonds, the true cost of borrowing $1 of private funds also includes the 5 cents of foregone tax revenue or a total of 10 cents. When properly calculated, then, according to the Water Resources Council (1971), the cost of federal borrowing gives about the same estimate of displacing $1 of private investment (10 percent in this example) as does private opportunity cost.

The Harvard School is named for the economics professors and students of Harvard University who recommend the use of the *social discount rate*, the rate reflecting the time preference of society for present consumption as opposed to waiting for future consumption. Society's rate of time preference can be thought of as the interest rate that consumers are willing to pay in order to consume now rather than waiting. It might range from 8 percent on home mortgages to 18 percent on automobile loans and a minimum estimate of the time preference rate is generally considered to be the rate accepted by the public on tax free government bonds (Howe, 1971). Since the social discount rate is sometimes construed to represent what society's time preference *should* be rather than the *actual* preference revealed by our actions, it has been argued that this rate may bear no relationship to interest rates paid in the money market or to rates earned on private investment (Howe, 1971; Barkley and Seckler, 1972). For a variety of reasons, most notably the 50 percent corporate income tax on private earnings and the imperfections of the market for borrowed funds, even the revealed time preference rate will not be the same as the opportunity cost rate. Although high time preference consumers push the average time preference rate up through 18 percent credit card purchases, the average private opportunity cost rate is also increased by the 20 percent corporate rates of return required to pay competitive after tax dividends (Howe, 1971). Thus the average time preference rate recommended by the Harvard School is less than the average private opportunity cost of the Chicago School.

The Water Resource Council (1971) approach will be described as an example of discount rate selection procedure. Although dated, this example still represents the philosophy underlying federal agency discount rate selection. The Water Resources Council "Proposed Principles and Standards for Planning Water and Related Land

Resources" begins as if the recommendations of the Chicago School are to be followed. The council initially states that the federal government should not displace funds in the private sector unless public return on investment at least equals that of private investment. The opportunity cost of federal investment is recognized as the real[1] rate of return on nonfederal investments. The best estimate of opportunity cost is described as the weighted average rate of return on private investment before income and property taxes and excluding the rate of general inflation.

The 10.4 percent weighted average rate of return on private investment reported by Stockfisch (1969) is cited as the best estimate available at that time of the federal opportunity cost. However, the council then discusses the "revealed preference" of the federal political process to transfer wealth to specific geographic regions of the country by subsidizing water projects (primarily by using an artificially low discount rate). Accepting income transfer as a legitimate federal objective, the "Standards" ultimately recommends that a discount rate of 7 percent be used in evaluating water projects during the five year period 1972–1976. As further justification of the 7 percent rate, there is brief mention of the long held notion that the federal discount rate should be based on the federal cost of borrowing. The argument is then made that correct calculation of borrowing cost includes foregone tax revenues on displaced private investment[2] as well as actual interest on federal bonds. The Council next states that since the full cost of federal borrowing is between 7 and 10 percent, the 7 percent recommended rate "approaches both the opportunity cost and the total cost of Federal borrowing."

In their final discount rate recommendations some two years later, however, the Water Resources Council (1973) did an about-face and specified that the discount rate should be based on the:

> ...average cost of Federal borrowing...during the 12 months preceding...on interest-bearing marketable securities...with remaining periods to maturity comparable to a 50 year period of investment: Provided however, that the rate shall be raised or lowered by no more than or less than one-half percentage point for any year.

The Water Resources Council (1973) recommended a discount rate of $6\frac{7}{8}$ percent for the remainder of 1974. Due to fears that a high discount rate would "foreclose opportunities which would otherwise be available when future water resource need of the nation become even more critical" (U.S. Congress, 1973), Congress overruled this Council recommendation with the Water Resources Development Act of 1974. The directives printed in the *Federal Register* (U.S. Congress, 1974) included the following:

> the interest rate...shall be based upon the average yield...on...securities...15 years or more remaining to maturity...in no event shall the rate be raised or lowered more than one-quarter of 1 percent for any year.

This policy, which is clearly based on the cost of borrowing rather than opportunity cost, resulted in discount rates of $5\frac{7}{8}$ percent for fiscal year 1974 (Hanke et al., 1975) and $6\frac{1}{8}$ percent for fiscal year 1976 (Bureau of Land Management, 1976).

Based on policy to date, it seems unlike that Congress will ever adopt either the opportunity cost rate of the Chicago School of the social discount rate of the Harvard School. Instead it will likely continue to pursue the middle ground of recommending a discount rate based on the interest rate paid on long term government bonds. The federal borrowing rate does provide an estimate of the time preference of the saving portion of society. But the federal borrowing rate does not include foregone tax revenue. This is clearly a serious omission, since most federal borrowing is not in the form of tax-free bonds.

Adjusting for Risk

Most range improvement projects involve some degree of risk. A drought following range seeding could cause the reseeding project to fail. Implementation of a planned grazing system may not bring as much improvement in range condition as was projected based on data from similar grazing systems. Despite the fact that the degree of risk for specific range improvement projects may be known in some detail (i.e., 1 out of 10 fall seedings of a certain species on a certain range site have failed in the past) risk is often ignored in benefit–cost analysis of public range improvement projects. The main justification underlying this omission is that from society's (or the individual taxpayer's) point of view, there is virtually no risk associated with a given federal project, since the chance of project failure is spread over all federal investments (and all taxpayers and federal bond purchasers). But there is a substantial difference between borrower's risk (the risk borne by the agency implementing the specific project) and lender's risk (the risk to the individual buyer of the federal bonds that are sold to finance the project). A plausible argument could probably be developed that the risk of failure should be incorporated in the benefit–cost analysis as a possible project cost. For obvious reasons, federal land administrators usually do not mention the possibility of project failure in their requests for project funds and, although in reality they represent the borrower, they have traditionally analyzed projects from the riskless viewpoint of the lender.

Howe (1971) pointed out that the portion of federal project benefits and costs accruing to private individuals does involve risk to these people and suggested a discounting procedure designed to take such risk into account. Howe proposed that the present value of the stream of project benefits (PVB) be calculated according to the following formula:

$$\text{PVB} = E(B_0) + \frac{E(B_1)}{(1 + r + r')} + \cdots + \frac{E(B_n)}{(1 + r + r')^n},$$

where $E(B_1)$ is the expected value of the benefits to be received at the end of year 1, n is the number of years the project is expected to produce benefits, r is the riskless discount rate, and r' is the risk (in percent) of the project not producing B benefits. Similarly, Howe recommends the following formula for discounting the cost stream:

$$\text{PVC} = E(C_0) + \frac{E(C_1)}{(1 + r - r')} + \cdots + \frac{E(C_n)}{(1 + r - r')^n}.$$

It should be noted that in the case of costs, the risk rate is subtracted from, rather than added to, the riskless rate as was suggested for benefits, thereby decreasing the amount that the costs are discounted. In explaining the negative sign of the risk rate for costs, Howe stated that "individuals avoid risk, not for fear of things turning out too well, but for fear of things turning out too poorly." Thus the risk is that one of the costs will be too high (requiring a subtraction from the riskless discount rate) and that the benefits will be too low (requiring that the benefits be discounted more heavily than for time alone).

Lind and Arrow (1970) suggest the use of information from the workings of the private market to calculate the risk premium that could be added to the ordinary federal discount rate to take the possibility of project failure into account. The rate of return being earned on similar investments on private land should approximate the risk adjusted discount rate $(r + r')$ listed above. Thus if the rate of return earned on a private rangeland reseeding is 12 percent, which supposedly includes an estimate of the risk associated with the project, and the risk-free discount rate specified for federal projects is 7 percent, a 5 percent adjustment could be added to account for risk. Definite improvements in efficiency of project selection might result from incorporating an adjustment for risk. As with other suggestions made in this chapter, however, the advantages would result only if the benefit–cost analyses of all federal projects were modified in this way and the first few public land administrators to adopt such a procedure would be at a distinct disadvantage in requesting improvement funds.

Adjusting for Inflation

As indicated by the Water Resources Council (1971) in specifying that the federal opportunity cost rate should be estimated in terms of the real rate of return on nonfederal investments, discounting calculations should account for the rise in the general price level. Although common in economic analyses of natural resource development and perhaps intended as a means of building a margin for error into economic evaluations, the combining of real prices (current prices held constant over the life of the project) with nominal interest rates (market rates that include a premium for expected inflation in addition to the inflation free opportunity cost rate) is technically incorrect.

Howe (1971) was on of the first authors to explicitly distinguish between nominal and real terms in specifying discount rates and estimating prices for benefit–cost analyses. He pointed out that in the absence of inflation, both prices and market derived discount rates would automatically be expressed in real terms. In this situation, which has not prevailed for a number of years in the United States, the following formula correctly calculated PNW in real terms:

$$\text{PNW} = -C_0 + \frac{(B_1 - C_1)}{(1 + R)} + \cdots + \frac{(B_n - C_n)}{(1 + R)^n},$$

where C_0 is initial project investment, B_n is the annual gross benefits for year n expressed at constant (real) prices prevailing at the time of investment, C_n is the annual costs (also at constant prices), and R is the discount rate that would be applicable if price levels remained constant. Howe then introduced inflation into the analysis and offers (without mathematical proofs) the following formula for calculating PNW in nominal terms:

$$\text{PNW} = -C_0 + \frac{(B_1 - C_1)(1 + I)}{(1 + R)(1 + I)} + \cdots + \frac{(B_n - C_n)(1 + I)^n}{(1 + R)^n(1 + I)^n},$$

where $(1 + I)$ is an inflationary term to adjust prices and the discount rate for I percent inflation per year. Since the inflationary terms cancel out, Howe concluded that:

> in the case of general inflation, it makes no difference whether we use (1) benefits and costs all stated in construction period (real) prices and a (real) discount rate containing no inflationary premium, or (2) benefits and costs in the prices of the year in which each is incurred and a discount factor that fully compensates for the rate of inflation.

Hanke et al. (1974) further developed the "either consistently real or consistently nominal" argument by providing a mathematical proof that the inflationary term used to correct the familiar discounting factor $(1 + i)^{-n}$ for general inflation results in a multiplicative rather than an additive effect. Their proof clearly demonstrates that the term $(1 + r)^{-n}$, where r is the nominal discount rate derived from market rates and accounting for both opportunity cost (or time preference) and inflation, is actually a product:

$$(1 + r) = (1 + R)(1 + I) \quad \text{or}$$
$$r = R + I + RI,$$

where R is the real (inflation-free) rate of interest and I is the annual rate of inflation, rather than simply the sum of R and I.

Hanke et al. (1975) also observed that both the final recommendations of the Water Resources Council (1973) and the official procedures established by the Congress (1974) in the Water Resources Development Act of 1974 incorrectly combine real prices with nominal discount rates. They make the point that since prices of benefits and costs were specified in real terms, the discount rates set by Congress and the Water Resources Council should have been real rather than nominal rates. Using the costs of borrowing approach upon which both recommendations were based, the correct real cost of borrowing would be about 4 percent (Hirshleifer and Shapiro, 1969), considerably less than the recommended $5\frac{7}{8}$ percent (Hanke et al., 1975). If the opportunity cost approach were used, the recommended real prices should be used in conjunction with the real opportunity cost of 10.4 percent (Stockfisch, 1969). Hanke also noted that the Office of Management and Budget (1972) did recommend a discount rate of 10 percent for federal project evaluation but that this recommendation was not adopted in the "water field" (in which range improvement projects are traditionally included).

Although published just prior to the work of Hanke et al. (1975) and not cited by these authors, a paper by Overton and Hunt (1974) took the nominal versus real consistency argument one step further by including a discounting procedures adjustment for increases in relative prices. They recommend that in place of the standard discounting term, $(1 + r)^{-n}$, the following term be used:

$$\left[\frac{(1 + r)}{(1 + I)(1 + V)} \right]^{-n},$$

where r is the nominal discount rate derived from the market, I is the annual rate of inflation, and V is the annual rate of increase in relative (real) price. If 6 percent were the nominal discount rate used in the standard discounting term, the inflation rate were 5 percent, and the rate of relative real price increase were 1.5 percent (as it has been for lumber products for a number of years), we can calculate what the implicitly assumed nominal discount rate would have to be in order for the standard discounting term to give the same results as the term recommended by Overton and Hunt. Substituting the above rates, setting the two formulas equal, and solving for r we obtain:

$$(1.06)^{-n} = \left[\frac{(1+r)}{(1.05)(1.015)} \right]^{-n},$$

$r = 12.97$ percent, which is not conservative even for a nominal discount rate. Clearly, the use of nominal discount rates in combination with real prices places a downward bias on the calculated present value of future costs or returns. Ignoring changes in real relative prices introduces a similar error. Due to the difficulty in forecasting future inflation it may be easier to convert nominal discount rates to real rates rather than to express real prices in nominal terms as recommended by Overton and Hunt (1974). No matter which conversion is preferred, when significant changes in either relative prices or the general price level are expected, estimates of such changes should be included in the project evaluation.

LITERATURE CITED

Arrow, K. J. and R. C. Lind. 1970. Uncertainty and the evaluation of public investment decisions. *Amer. Econ. Rev.* 60.

Barkley, P. W. and D. W. Seckler. 1972. *Economic Growth and Environmental Decay—The Solution Becomes the Problem.* Harcourt Brace Jovanovich, New York. 193 pp.

Bromley, D. W. 1981. The role of economic analysis in public range management: A discussant paper. Paper presented at the *Workshop on Applying Socioeconomic Techniques to Range Management Decision Making.* Boise, Idaho, May 11-12. National Research Council.

Brown, W. G., A. Singh, and E. Castle. 1964. *An Economic Evaluation of the Oregon Salmon and Steelhead Sport Fishery.* Oregon Agricultural Experiment Station Technical Bulletin No. 78.

Bureau of Land Management. 1976. *Allotment Management Plan (AMP) Economic Analysis.* U.S. Department of the Interior Instruction Memorandum No. 76-455, Washington, DC.

Bureau of Land Management. 1979a. Public Lands and Resources; Planning, Programming, and Budgeting. Circular No. 2451. 10 pp.

Bureau of Land Management. 1979b. BLM Instruction Memorandum No. 80-571. 100 + pp.

Bureau of Land Management. Undated. Social and Economic Analysis. Policy and Action Plan. 56 pp.

Bureau of the Budget. 1956. Circular No. A-47. Washington, DC.

Clawson, M. 1959. Methods for measuring the demand for and the value of outdoor recreation. Resources for the Future Reprint No. 10, Washington, DC.

Committee on Risk and Decision Making. 1982. Excerpts from Academy Reports: On Uncertainties in Assessment of Risks. *National Academy of Sciences News Report* 37:18–21.

Council for Agricultural Sciences and Technology. 1974. Livestock grazing on federal lands in the 11 western states. *J. Range Manage.* 27:174–181.

Dickerman, A. R. and W. E. Martin. 1967. *Organization, Costs, and Returns for Arizona Cattle Ranches.* Department of Agricultural Economics File Rep. 67-6. University of Arizona, Tucson.

Dwyer, J. F., J. R. Kelly, and M. D. Bowes. 1977. Improved procedures for valuation of the contribution of recreation to national economic development. Water Resources Center Report No. 128, University of Illinois, Urbana-Champaign.

Dyer, A. A. 1981. Public natural resource management and valuation of nonmarket outputs. Paper presented at the *Workshop on Applying Socioeconomic Techniques to Range Management Decision Making. Boise, Idaho, May 11–12.* National Research Council.

Fulcher, Glen D. 1973. Grazing systems—A least cost alternative to proper management of the public lands. Abstracts of Papers of *26th Annual Meeting, Society for Range Management, Boise, Idaho.* Society for Range Management. Pp. 21–22.

Gardner, B. D. 1981. The role of economic analysis in public range management. Paper presented at the *Workshop on Applying Socioeconomic Techniques to Range Management Decision Making. Boise, Idaho, May 11–12.* National Research Council.

Gardner, B. D. and A. LeBaron. 1965. Lectures on economics of water resource development and conservation. Lectures presented at the *Summer Institute in Water Resources, June 21–August 13, 1965.* Utah State University, Logan. 144 pp. (Mimeo.).

Garrett, J. R., G. J. Pon, and D. J. Arosteguy. 1970. *Economics of Big Game Resource Use in Nevada.* Nevada Agricultural Experiment Station Bulletin No. 25. 22 pp.

Godfrey, E. B. 1983. Economics and multiple use management of federal lands. *Proceedings of Range Economics Symposium and Workshop, August 31–September 2, 1982, Salt Lake City, Utah.* U.S. Forest Service General Technical Report No. INT-149. Pp. 77–81.

Gushee, C. H. 1968. *Financial Compound Interest and Annuity Tables.* Financial Publishing Co., Boston, MA. 884 pp.

Hanke, S. H., P. H. Carver, and P. Bugg. 1975. Project evaluation during inflation. *Water Resources Research* 11:511–514.

Herfindahl, O. C. and A. V. Kneese. 1974. *Economic Theory of Natural Resources.* Merrill, Columbus, OH. 405 pp.

Hirshleifer, J. and D. L. Shapiro. 1969. The treatment of risk and uncertainty. In *The Analysis and Evaluation of Public Expenditures: The PPB System,* vol. 1. U.S. Government Printing Office, Washington, DC. Pp. 505–530.

Hotelling, H. 1949. The economics of public recreation. Prewit Report, Washington, DC. Unpaged.

Howe, C. W. 1971. *Benefit–Cost Analysis for Water Systems Planning.* American Geophysical Union Water Resources Monograph No. 2. Publications Press, Inc., Baltimore, MD. 144 pp.

Howitt, R. E. 1981. Public natural resource management and valuation of nonmarket outputs: A discussant paper. Paper presented at the *Workshop on Applying Socioeconomic Techniques to Range Management Decision Making. Boise, Idaho,*

May 11–12. National Research Council.

Knetsch, J. L. and R. K. Davis. 1974. Comparisons of methods for recreation evaluation. In *Land and Leisure: Concepts and Methods in Outdoor Recreation.* D. W. Fisher, J. E. Lewis, and G. B. Priddle (Eds.). Maaroufa Press, Chicago. Pp. 175–191.

Marshall, A. 1947. *Principles of Economics*. MacMillan, London. P. 124.

Martin, W. E. and G. L. Jefferies. 1966. Relating ranch prices and grazing permit values to ranch productivity. *J. Farm Econ*. 48:223–242.

Mishan, E. J. 1976. *Cost–Benefit Analysis*. Praeger, New York. 454 pp.

National Park Service. 1950. A method of evaluating recreation benefits of water control projects. U.S. Department of the Interior, Washington, DC.

O'Connell, P. F. and R. S. Boster. 1974. Demands on national forests require coordinated planning. *Ariz. Rev.* 23:1–7.

Overton, W. S. and L. M. Hunt. 1974. A view of current forest policy, with questions regarding the future state of forests and criteria of management. *Transactions of 39th North American Wildlife and Natural Resources Conference*. Wildlife Management Institute, Washington, DC.

Region Research Project W-79. 1968. Economic analysis of ranch and range management decisions on western livestock ranches. Western Agricultural Experiment Station, Economics Research Service, and Cooperative State Research Service USDA.

Row, C., H. F. Kaiser, and J. Sessions 1981. Discount rate for long-term forest service investments. *J. Forestry* 79:367–376.

Schuster, E. G. and J. G. Jones. 1983. Extramarket valuation for resource allocation: A critique. *Proceedings of Range Economics Symposium and Workshop, August 31–September 2, 1982. Salt Lake City, Utah.* U.S. Forestry Service General Technical Report No. INT-149. Pp. 63–73.

Smith, A. H. and W. E. Martin. 1972. Socioeconomic behavior of cattle ranchers with implications for rural community development in the West. *Am. J. Agric. Econ.* 54:1217–225.

Smith, S. C. and E. M. Castle. 1964. *Economics and Public Policy in Water Resource Development*. Iowa State University Press, Ames, IA. 463 pp.

Stockfisch, J. A. 1969. Measuring the opportunity cost of government investment. Institute for Defense Analysis Research Paper No. P-140, Arlington, VA. 28 pp.

U.S. 87th Congress. 1962. *Senate Document 97: Policies, Standards, and Procedures in the Formulation, Evaluation, and Review of Plans for Use and Development of Water and Related Land Resources*. U.S. Government Printing Office, Washington, DC.

U.S. Congress. 1968. *Federal Register*. December 24, 1968. P. 19170.

U.S. Congress. 1970. *The National Environmental Policy Act of 1969*. Public Law 91-190 (42 U.S.C. 4321–4347).

U.S. Congress. 1974. *Water Resources Development Act of 1974*. Public Law 93-251, Section 80, p. 23.

U.S. Congress. 1974. *Federal Register* 39:29242–29243.

U.S. Forest Service. 1973. *Pinyon-Juniper Chaining Program on National Forest Lands in the State of Utah*. Final Environmental Statement, Intermountain Region, Ogden, Utah.

U.S. Forest Service. 1979. National Forest System Land and Resources Management Planning. *Federal Register* 44:53928–53999.

U.S. Forest Service. 1980a. *Economic Analysis of Allotment Management Plans and Range Improvements. WO-Range.* Review Draft No. FSH 2209.11. 88 pp.

U.S. Forest Service. 1980b. *Economic Analysis for Forest Planning.* Draft No. FSH 1909.12. 53 pp.

U.S. Interagency Committee on Water Resources. 1950. *Proposed Practices for Economic Analysis of River Basin Projects,* Washington, DC. (Revised 1958.)

U.S. Office of Management and Budget. 1972. *Discount Rates To Be Used in Calculating Time Distributed Costs and Benefits.* Office of Management and Budget Circular No. A-94, Washington, DC.

U.S. Water Resources Council. 1971. Proposed Principles and Standards for Planning Water and Related Land Resources. *Federal Register* 36:24144–24194.

Water Resources Council. 1973. *Federal Register* 38:174.

Wennergren, E. B. 1964. *Demand Estimates and Resource Values for Resident Deer Hunting in Utah.* Utah Agricultural Experiment Station Bulletin No. 496. 44 pp.

Workman, J. P. 1975. Wildlife and recreation on U.S. rangelands—The economic aspects. In *Arid Shrublands, Proceedings of the Third Workshop of the United States/Australia Rangelands Panel. April 4, 1973, Tucson, Arizona.* Pp. 131–134.

Workman, J. P. and S. K. Fairfax. 1981. Applying socioeconomic techniques to range management decision making. In *Developing Strategies for Rangeland Management.* Committee on Developing Strategies for Rangeland Management, National Research Council, Washington, DC. Chapter 5, pp. 75–88.

Workman, J. P. and J. E. Keith. 1975. Economics of soil treatments in the upper Colorado. *Watershed Management Symposium Proceedings. August 13, 1975, Logan, Utah.* American Society of Civil Engineers. Pp. 591–596.

Workman, J. P. and C. R. Kienast. 1975. Pinyon-Juniper manipulation—Some socioeconomic consideration. *The Pinyon-Juniper Ecosystem—A Symposium. May 2, 1975, Utah State University, Logan, Utah.* Pp. 163–177.

TEXT NOTES

1 The present value of $1 to be received annually for 200 years discounted at seven percent interest is $14.283 (Gushee, 1968).

2 The term *cost effectiveness analysis,* as used here is defined according to standard economics terminology. Cost effectiveness analysis is used to select the minimum cost approach of accomplishing a goal that has already been set. This traditional definition should not be confused with the incorrect use of the term "cost effectiveness" by the U.S. Forest Service (1980a) as a synonym for the concept of economic efficiency described in Chapter 9. Care should also be taken to not be misled by the term "cost efficient" that BLM (undated) sometimes uses to mean "cost effective" but that the U.S. Forest Service (1979) uses as a synonym for economically efficient.

3 This reasoning has sometimes, and perhaps unnecessarily, been taken one step further in cost effectiveness analysis. It has been argued that, since it was imperative to achieve the goal by some means, the net benefits accruing to the project selected can be represented by the difference in cost between the next least expensive method available and the least-cost method actually chosen (Howe, 1971).

4 In practice, even this simple (but logical) cost effectiveness rule of selecting the least-cost method from those promising identical benefits is not always followed. The U.S. Forest Service (1973), for example, reported the following benefit–cost ratios for three types of pinyon-juniper control: 1.08 for chaining, 1.30 for individ-

ual tree removal, and 1.63 for prescribed burning. Although all three treatments were expected to produce identical benefits and despite chaining being the most expensive method considered, chaining was the method selected for use in Utah. Currently, the U.S. Forest Service (1980b) would likely describe each of these three methods as (economically) feasible but would identify burning as the most "cost efficient" practice.

5 If the logic is carried a bit further, such an approach does provide some useful information. If a decision is made to not cut the tree, then the value of the timber not harvested represents what the nesting tree *must* be worth in order for this to be a rational decision. Thus, the assumption underlying the manager's decision is at least made explicit.

6 Use of consumer surplus in nonmarket product valuation is primarily an attempt to avoid overestimates and underestimates of value as calculated by the gross expenditure method. Actual expenditure ($300 at the $1 price) is often small for public land nonmarket benefits compared with willingness to pay. When first introduced to the consumer surplus technique, students often pose two questions: Why treat only the surplus as the value of the commodity? Why not treat the entire amount consumers are willing to pay (the actual expenditure plus consumer surplus) as the value? Wennergren (1965) provides an answer to these questions in pointing out that the consumer surplus concept is logically consistent with real estate appraisal techniques of valuing land and residential and commercial properties. Consumer surplus as a measure of net value is somewhat analogous to the appraisal technique of capitalizing net income in estimating property values.

7 Reprinted by permission from *Economics and Public Policy in Water Resource Development* by S. C. Smith and E. M. Castle © 1964 by the Iowa State University Press, Ames, Iowa 50010.

APPENDIX I NOTES

1 The fact that the real rate of return (nominal rate of return minus the rate of general inflation) was specifically recommended in the original standards printed in the *Federal Register* should be borne in mind when reading the Adjusting for Inflation section to follow.

2 It should be noted that this argument is a sharp departure from previous official federal policy concerning discount rate selection. Congress (1968) specified that discount rates were to be based on "the yield during the preceding fiscal year on interest-bearing marketable securities of the United States which at the time the computation is made have 15 years or more remaining to maturity."

INDEX

Aftermath, 151
Alternative products, price of, 32
American Journal of Agricultural Economics, 5
Animal unit, 149
    equivalent, 149
    month, 146
Annual deposit to yield one, 134
Annual percentage rate (*see* Interest, simple annual)
Annuity, 125
Appreciation, land, 19
Assets
    current, 21
    fixed, 21
    working, 21
Average cost curve, 36
    long run, 72
    short run, 71

Benefit–cost ratio, 142, 198
Best sites first, principle of, 79
Borrowing cost, 119
    real, 131
    nominal, 132
Break-even price (*see* Minimum acceptable price)
Budgeting, 155
    partial, 154, 155
Burning, 145
Buyers, number of, 32

Calf crop percentage, 14
Capital, 51
    budgeting, 164
    constraint, 51
    unlimited, 51, 79
Capitalization, 131
    rate, 131
        market, 131
Capitalized value, 196
Carrying capacity, 144, 152
Circular A-47, 184
Committee on Economics of Range Use and Development, 4
Comparative economic advantage, law of, 12
Competitive market, 35, 50, 97
    purely, 97
Complements, product, 32
Compounding, 120
Computer techniques, 156
Confinement feeding, 152
Corner solution, 106

Cost economies (*see* Economies of size)
Cost effectiveness, 190
Cost surface, 144
Costs
   maintenance, 142
Cow to bull ratio, 150
Credit decisions, 138
Critical projects, 11
Cull bulls, 151
Cull cows, 150
Curves
   asymptotic, 45
   concave, 45
   convex, 45
   sigmoid, 45

Demand, 25, 69
   curve, 27
   curve shift, 30
   curve shifters, 32
   excess, 30
   law of, 26
   law of downward sloping, 27
   movement along curve, 31
   schedule, 25
Depreciation, 15
Diminishing marginal productivity, law of, 42
Diminishing returns, law of, 40, 42, 51, 86
Discount rate, 198
   adjusting for inflation, 204
   adjusting for risk, 203
   borrowing, 202
   importance of, 198
   nominal, 204
   opportunity cost, 201
   real, 200
   selection of, 200
   social, 201
   time preference, 201
Discounting, 118, 124
Down payment, 18, 135

Economic efficiency, 183
Economic equity, 183
Economic Research Service, 4
Economics, 1
   agricultural, 2
   objective, 183
   production, 2
   ranch, 2
   range, 1–5
   resource, 2
   subjective, 183
Economically feasible, 11, 165, 186
Economies of size, 144
Elasticity
   coefficient, 34
   elastic demand, 34
   inelastic demand, 34
   price elasticity of demand, 34
   quantity coefficient, 34
   unitary, 34
Equilibrium, 29, 69
   long run, 72
   price, 29, 69
   quantity, 29
Equating at the margin, 51, 61, 64, 67, 68
Equity (*see* Owned capital)
Expansion path, 88, 95
Expectations, consumer, 32

Factors of production
   fixed, 40
   variable, 40
Federal income tax, 133
Financial statement, 20
Firm size, 71
Fixed cost
   average, 50
   total, 50
Fixed factor, 50
Flood Control Act of 1936, 184
Flow, 124
   cash, 124
   costs, 125
   future, 130
   perpetual, 131
   returns, 125
   resource, 12
Forage, 145, 147
   allowable use of, 151
   balance, 147
      chart, 149
      seasonal, 148
   requirements, 146
   seasonal surplus, 147
   usable, 145
   utilization constraint, 149
   value as lease, 147
   value for livestock, 147
Future, 118
   costs, 118
   returns, 118
   value, 121
Future worth of annual deposits, 135
Future worth of one, 121

Gains
   per acre, 55
      maximum, 55
   per animal, 55
      maximum, 55
Grazing deferment, 157, 175, 187
   costs of, 157
Grazing fees, 191
   federal, 191
   private, 147
Grazing leases, 61, 147
Green Book, 184

INDEX

Herbage, 145
Herd complement, 153
Herd size, 144
Income, 17
  to capital, 17
  disposable, 32
  to labor, 17

Induced costs, 175
Inflation, 119
  rate, 132
Inflection point, 46
Input, 48
  adjustments for optimum, 96
  fixed, 48
  optimum combination of, 85
    maximum product, 88
    minimum cost, 87
  price, 32
  variable, 48
    intensification of, 71
Interest, 18
  compound, 120
  nominal, 121
  payment, 18
  real, 119, 141
  simple annual, 138
Interest rate
  real, 158
Internal rate of return, 143, 198
Inverse price line (*see* Isorevenue)
Investment, 141
  criteria, 142, 163
    disagreement among, 164
    inconsistent, 166
  initial, 141
Isocline, 93
Isocost, 87
  line, 54
Isoquant, 86
  characteristics, 91
  map, 89
Isorevenue, 109
Iterative calculations, 175

Journal of Farm Economics, 4
Journal of Range Management, 4, 5

Land appreciation, 132
Land use, 102
  alternatives, 103
  decisions, 102
  interactions, 103
Least cost alternative, 192
Liabilities
  current, 21
  fixed, 21
  working, 21
Life of project, 130

Limited resources, 11
Limiting season, 152
Linear programming, 147, 155
Living expenses, net return available for, 18
Loan, 18
  add-on method, 138
  discount method, 138
  service, 18
Long-run, 69, 71
Loss, 66
  economic, 66

Marginal costs, 27
  increasing, 27
Marginal factor cost, 73
  curve, 74
Marginal physical product, 107
Marginal rate of substitution, 107
  constant, 103
  decreasing, 115
  increasing, 106
Marginal rate of technical substitution, 91
Marginal revenue, 50
  line, 66
Marginal utility, 26
  decreasing, 26
Marginality, 48
  general model of, 61
  principle of, 48
Market equilibrium, 29
Market price, 102
Market priced benefits and costs, 187
Maximum income, 10
Metabolic requirement ratio, 150
Minimum acceptable price, 68
  short-run, 68
  long-run, 69
Minimum loss, 67
Mortgage rate, nominal, 20
Multiple use, 102, 116
Mutually exclusive projects, 164, 173

National Environmental Policy Act, 184
Negative return, 163
  on ranch capital, 163
Net present value (*see* Present net worth)
Net revenue, 36, 51
  maximum, 52
Net worth, 21
Nonmarket priced benefits and costs, 187
Nonmarket pricing, 111, 190
  consumer surplus, 194
  cost of alternatives, 194
  gross expenditure, 194
  production cost, 193
  willingness to pay, 194
Nonquantifiable benefits and costs, 187
Nonrenewable resource, 12
Nonuniform flows, 174

Normalization, 167
  of benefit–cost ratio, 170
  for differences in project life, 170
  for differences in project size, 167
  of internal rate of return, 170

Operator labor, value of, 17, 20
Opportunity cost, 18, 119, 193
  real, 141
Optimum, 75, 96
  allocation between enterprises, 75
  combination of investments, 172
  combination of outputs, 102
  constrained, 77
  input, 48, 74
  output, 48, 56
    necessary conditions for, 96
    sufficient conditions for, 96
  scale, 48
Owned capital, 20
  net proceeds to, 20
  rate of return on, 20

Perquisities, 17
Physical product, 40
  average, 41
  marginal, 41
  total, 40
Planning horizon, 11
Pregnancy evaluation, 150
Present net worth, 142, 198
Present value, 121, 141
  of a flow, 141
Present worth of one, 126
Present worth of one per period, 128
Price, 48
  fixed input, 48
  net product, 109
  nominal, 121
  product, 48
  variable input, 48
Price line, 54
  inverse, 54
Price-payer, 71
Price ratio, 54
  inverse, 54
Price-taker, 71
Principal, 18–19, 131
  balance, 18
  payment, 19
Production
  economics, 38
  efficiency of, 42
  function, 39, 48, 103
    graphical, 40
    tabular, 40
  possibility curve (*see* Transformation curve)
  stages, 65

Products, 103
  antagonistic, 114
  competitive, 103
  complementary, 110
  supplementary, 112
Profit, 56, 66, 142
  economic, 66
  factors affecting, 56
  maximum, 51, 63, 67–68
    constraints on, 82
  negative, 66
Project analysis, 184
Project justification, 185
Project life, 143
Public lands, 183
Public ranges, 183
Pure profit (*see* Economic profit)

Quantifiable benefits and costs, 187

Ranch income, net, 15
Ranch income statement, 15
  standard, 15
  modified, 18
Ranch size, 144
Range, 1
  condition, 145, 161
  deterioration, 52
  improvements, 141
  land, 1
  management, 1, 3
  readiness, 148
  site, 145
    selection for treatment, 160
Ranking of investments, 172
Rate of return, 13, 130, 163–164
  on added investment, 163
  internal, 130
  on investment, 13
  nominal, 20
Reinvasion, 156
  of shrubs, 156
Reinvestment of project returns, 167
Renewable resource, 12
Replacement bulls, 151
Replacement heifers, 150
Resources, 102
  fixed, 102
  variable, 102
  surplus, 113
Return, 109
  annual net, 141
  maximum net, 109
  nominal, 141
  real, 141
Revenue, 50
  gross, 35
  marginal, 50
  total, 50

INDEX

Ridge lines, 92
Risk, 119, 203
  aversion, 51
  borrower's, 203
  lender's, 203

Scale, 72
  cattle ranch economies of, 73
  diseconomies of, 72
  economies of, 72
  operation, 71
  returns to, 73
Senate Document No. 97, 184
Short run, 50, 68
Shortcut method, 54
Simultaneous solution, 100
  of multiple equations, 100
Single use, 116
Sinking fund, 133
Soil Conservation Service, 145
Stages of production, 41
  relevant stage, 42
Stock, 125
  cost, 125
  return, 125
Stock count chart, 150
Stock resource, 12
Stocking rate, 48
  adjustments to price changes, 56
  biological optimum, 52
  economic optimum, 52
Subcommittee on Range Research Methods of the Agricultural Board, 5
Substitutes, product, 32
Suppliers, 32
Supply, 27
  curve, 28, 69
  curve shift, 31
  curve shifters, 32
  movements along curve, 32
  schedule, 27
  shortage, 30
  surplus, 30
Sustained yield, 10, 52, 109, 145
  maximum, 10, 52

Tangency, 54, 88
  isocost and isoquant, 88
  isocost and total product, 54
  isorevenue and transformation curve, 109
Tastes, consumer, 32
Technology, 32
  advance, 32
Time, effects of, 118
Time preference, 119, 186
Total cost, 51
  average, 51
  curve, 63
  marginal, 51
Total digestible nutrients, 148
Total fixed cost curve, 63
Total physical product, 48
  curve, 62
Total value product, 73
  curve, 74
Total variable cost curve, 63
Trade-off, 11 (*see also* Transformation curve)
Tragedy of the commons, 52
Transformation curve, 103
  convex, 116
Truth-in-lending, 138

Value marginal product, 73, 110
  curve, 74
Variable cost, 50
  average, 51
  total, 51
Variable input, 59
  conversion to a fixed input, 59
Variable proportions, law of, 43
Vegetation guide, 145

Water Resources, Council, 184
Weaner calves, 154
Weather, 32
Western Agricultural Economics Association, 4
Western Agricultural Economics Research Council, 4
Western Journal of Agricultural Economics, 5
Willingness
  to buy, 29
  to sell, 29